BUILDING THE KING'S HIGHWAY

BUILDING THE KING'S HIGHWAY

LABOR, SOCIETY, AND FAMILY ON MEXICO'S *CAMINOS REALES*, 1757–1804

Bruce A. Castleman

THE UNIVERSITY OF ARIZONA PRESS
TUCSON

The University of Arizona Press
© 2005 The Arizona Board of Regents
All rights reserved

♾ This book is printed on acid-free, archival-quality paper.
Manufactured in the United States of America
10   09   08   07   06   05   6   5   4   3   2   1

Library of Congress Cataloging-in-Publication Data

Castleman, Bruce A.
  Building the King's Highway : labor, society, and family on Mexico's
  Caminos Reales, 1757–1804 / Bruce A. Castleman.
     p. cm.
  Includes bibliographical references and index.
  ISBN 0-8165-2439-4 (cloth : alk. paper)
  1. Road construction workers—Mexico—History—18th century.
2. Roads—Mexico—History—18th century. 3. Labor market—Mexico—
History—18th century. 4. Mexico—History—Spanish colony, 1540–
1810. 5. Mexico—Social conditions—18th century. I. Title.
HD8039.R62M632 2005
331.7'6257'0972409033—dc22
                                                        2004023859

To Penny Castleman

CONTENTS

Figures and Tables

Acknowledgments

The debts and thanks I owe for this project are many and far-flung. Without the efforts of the staff of the Archivo General de la Nación in Mexico City, this research project would never have materialized. They kept up a steady flow of documents during my visits and provided high-quality microfilm that enabled me to continue my archival research far beyond the time I could spend in Mexico City. The crew of Galería 4 tirelessly brought me book after book of colonial documents. The director of the Archivo General de Indias in Seville, doña Pilar Lázaro de la Escosura, and her entire staff were also greatly helpful and cooperative during both of my visits there, in particular Sra. Socorro Prous, who helped me cybernavigate the newly computerized sections of documents during my first trip. At the Archivo Municipal de Orizaba, Lic. Rita Villalobos Pereyra spared no effort on my behalf. Soldiers assigned to the offices of the Servicio Histórico Militar in Madrid went out of their way to help me get what I needed. Closer to home, I always received cheerful support from University of California librarians at the Tomás Rivera Library in Riverside, the Geisel Library in San Diego, and the Bancroft Library in Berkeley. I must make special mention of the cooperative Interlibrary Loan staff of the San Diego State University Library.

Robert Patch has been an unfailing source of support, advice, and helpful criticism from the project's inception. Eric Van Young's unselfish assistance put me back on track time and again. Many other friends and colleagues contributed their insights and assistance and offered encouragement over the years. They are too numerous to mention all of them here, but I wish to single out Matthew Restall, Iris Engstrand, Roger Ransom, Carlos Cortés, Charles Wetherell, Robert McCaa, William B. Taylor, Joseph P. Sánchez, Erick Langer, Beth Pollard, Heike Schmidt, Farid Mahdavi, Luis Murillo, Brian Loveman, and Tom Davies for their help in so many different ways as I wended my way along this scholarly journey. The anonymous reviewers of the University of Arizona Press provided invaluable recommendations. And I thank G. Pope Atkins in particular for introducing me to the study of Latin America when I was an undergraduate. Whatever is good about this study is owed to those who helped me along the path, and I thank them profusely. They bear no responsibility for any deficiencies. The latter stem from my own shortcomings, which even they could not overcome.

Portions of chapter 2 and chapter 5 previously appeared in modified forms in the *Colonial Latin American Historical Review* and the *Colonial Latin American Review*, respectively. I gratefully acknowledge each journal's kind permission in allowing the use of those articles in this book.

I could never have written this book without the support of my family. My parents taught me to love learning from my youngest days, even if my own interests differed from what the schools wanted me to study. My sons, David and Charles, sacrificed a great deal of quality time over the years while I worked on the dissertation for which much of the research was originally conducted. But most important of all, I thank my best friend for everything. Penny Castleman, my wife, has been the rock upon whom I have leaned for years. No sooner did my years of going to sea end than she had to put up with my self-sequestrations for study, research, and writing. Her smiles are infinite, her good-humored patience is endless, and there is not enough ink in the world to give her the credit she deserves.

CHAPTER ONE

INTRODUCTION

The kings of Spain ruled a globe-girdling empire for centuries. Effective colonialism depended in great part on lines of communication that ran both over land and over the seas. From Mexico City, in a geographic sense the center of the Spanish Empire, roads connected Atlantic and Pacific ports, and others emanated hundreds of miles to an isolated northern frontier dotted by Spanish presidios and missions. Through these tentacles, the king of Spain established his sovereignty, but the roads also served a variety of other purposes.

Silver bullion moved from Mexican and Peruvian mines to Spain and also from Mexico to Manila, much of it eventually absorbed into the markets of China. An overland route connecting Acapulco to Mexico City and Veracruz linked transpacific Manila galleons to American treasure fleets bound for Spain. Along this avenue of trade, silver was exchanged for precious Asian luxury goods, which were carried back to Acapulco, Mexico City, and Seville.

The importance of the silver trade to the Spanish colonial effort has been recognized for years and continues to draw historians' attention today. Some contend that Chinese demand for silver was a crucial factor in Spain's ascendance to world power, whereas others draw particular attention to Spain's role in creating and expanding the transatlantic economic system during the sixteenth through the eighteenth centuries.[1] The overland road links in this world network, however, also served local markets and local persons. The struggles over road policy between the Bourbon colonial state, the important *consulados* (guilds) of Mexico City and Veracruz, and more localized interests are analyzed herein. As will be seen, the Consulado de México triumphed because it achieved its real objective by connecting the capital city with Toluca to the west and with Puebla to the east. These comparatively high-level issues are both interesting and informative in their own right, but here they also fill in the otherwise rather empty backdrop behind the construction workers themselves, who are the central figures of this particular story.

This book is a social history of road-construction laborers, somewhat broadly defined, in late Bourbon Mexico. The core archival documentation consists of employment and expense records from construction and maintenance efforts as well as census records from the late eighteenth century. During the time period of this study, there occurred a major shift in the methods used by the Spanish colonial regime to mobilize the supply of unskilled labor in central Mexico. During the 1760s, a free-wage labor regime replaced a draft-labor system that had been brokered through the leaders of colonized indigenous communities. All the while, artisan workers remained under a free-wage regime, but some found their wages reduced to the unskilled level, essentially pushing them into those ranks. With wage reductions in a time of inflation toward the end of the century, these workers' situations deteriorated over fifty-odd years. The decades of the late eighteenth century thus constitute an important conjuncture in time marking fundamental changes in the nature of colonial coercion of the Mexican laboring persons and, as is also shown, changes in the largely symbolic identities they asserted for themselves.

This study advances our understanding of labor in late-colonial Latin America through its analyses of wages actually paid to named individuals from one week to the next. A valuable corpus of literature has already established the series of wage rates paid to silver miners and to workers on public-works projects as well as on the Bourbon state's monopoly enterprises, such as tobacco production. These studies in turn complement the insights into the rural labor market offered by the various studies of haciendas during the late-colonial period. None of these analyses was able to incorporate the sort of weekly wage rosters used here simply because so few of the rosters have survived in the archives and therefore did not exist for many successful studies. The basic approach to using the weekly employment rosters is informed by Tamara Hareven's study of Amoskeag, New Hampshire, a twentieth-century mill town.[2] The Mexican rosters I used here are important because they corroborate the findings of many earlier studies with analyses of working people's actual participation in an aspect of the colonial-era labor market.

Although a *longue durée* is well beyond the time period of this study, the structure I used is nonetheless shaped by concepts advanced by Fernand Braudel and other historians of the Annales school. Braudel's *Civilization and Capitalism, 15th–18th Century* provides the basic framework, but the order of analysis of his three volumes is reversed here. Thus, the "perspective of the world" sets the broadest context, the world trade network sketched earlier. I next proceed to "wheels of commerce" in chapter 2, which presents another perspective on the highway network in late-eighteenth-century Mexico, or, as it was known at the time, the Viceroyalty of New Spain. A rough and patchy system was adequate to meet the goals of the Spanish extractive empire and to allow the shipment of high-value, low-volume goods such as silver bullion and Asian luxuries. But to support even that very valuable trickle, a large population with an attendant infrastructure of local markets was required. The colonial regime in Mexico was greatly concerned to improve local and regional commerce with better roads, so in chapter 2 I move from global concerns to local ones.

The next two chapters focus on those local concerns—or, to use Braudel's title, on the "structures of everyday life." Chapter 3 analyzes unskilled laborers on road-construction and repair projects at the point of production. Changes in the structure of the labor market are perhaps most visible here, as the colonial regime ceased to rely on a supply of drafted indigenous labor for unskilled work. To show the eighteenth-century Mexican situation in a broader historical context, I make comparisons with corvée labor systems in different places and at different times in history, including ancient Rome, pre-Conquest Peru, the Ottoman Empire, and imperial China. Chapter 4 applies the same analysis to skilled laborers on those four projects, draws comparisons with unskilled laborers, and allows some conclusions about the labor market in late-eighteenth-century Mexico.

History is more than the social and economic processes highlighted here. History is about people. People were part of those processes, true enough, but the people in this story were more than mere ants working on projects. Along the way, the reader will meet many individuals of eighteenth-century Mexico, just as I met them while researching the archives for the traces of themselves that they left behind. Some of them are important historical figures, such as don Antonio Primo de Rivera. Those familiar with Spanish history will immediately recognize the surname. A Primo de Rivera surrendered Manila to U.S. forces in the War of 1898. Another ruled as dictator under the reign of Alfonso XIII, and that man's son founded the Falange, the Spanish fascist party. An important man in eighteenth-century Mexico and an important family. The reader will also meet different viceroys and lesser-known members of elite colonial society, such as don Marcos González, an influential tobacco planter from Orizaba, and Captain don Manuel Agustín Mascaró, a military engineer who oversaw many important construction projects, including the highway from Mexico City to Toluca that looms large in this study.

Historians do much of their research by tracking documents filed in archives such as those in Mexico City and Seville or somehow retained in other places. People such as González, Mascaró, and Primo de Rivera left behind sizeable document trails that are readily followed. They speak clearly to the future that followed them, and in the case of don Marcos, he was probably yelling at the top of his lungs the whole time. But what about the tradesmen and laborers whose hands actually built these highways and bridges? Those people offer only the faintest of whispers for us. How then to hear from the nearly silent? One way for social historians to amplify those whispers is through prospography, or collective biography. Census manuscripts, parish records, and employment rosters often provide the means of amplification. Even though the results may not fit exactly this particular individual or that one, analysis of trends in such long lists of people gives an idea of what a society as a whole might have been like.

Chapter 5 fleshes out something of the lives of those working persons. There, they become "real people," no longer merely names on weekly employment records. This study is a history of people who built roads much more than it is a history of road construction or of roads themselves. As befits a social history, it also considers these persons' existences away from the point of production. This chapter looks only at persons in Orizaba, a result of the nature of the extant documentation. Of the four sets of construction records I reviewed, only the one from Orizaba's Puente de Escamela is a complete set. Finding it was thus like an archaeologist's discovering a complete example of pottery that makes sense of other broken pieces: the Orizaba records greatly help in making sense of the fragments surviving from the Xalapa repairs and from the Mexico City–Toluca highway construction. Serendipitously (to say the least), a major census was taken at the same time as the Puente de Escamela was constructed, and some names can be linked between the sets of data. Equally important, one of the few manuscripts surviving from an earlier census is also from Orizaba. Linking census manuscripts with employment records enables the use of the methods of historical demography to examine a host of social indicators, including marriage preference, family structure, and differences over time between the intent and the result of how the *sistema de castas* was used to classify persons according to their degrees of African, American, and European ancestry.

This study considers persons who lived in a colonial empire. Students of colonial situations are frequently tempted to attribute everything that happens in these situations to the forces of "colonialism," in particular those things that by nature seem oppressive. They often do so without making comparisons with "noncolonial" situations and in the process do little to expand or refine our knowledge of colonialism itself. At several points in this study, I draw comparisons with other secondary studies of parallel practices from eighteenth-century Europe. These comparisons are but a small initial step on a long path toward the development of comparative histories of colonialism itself. This study lends itself well to such an ambitious enterprise because road building and improvements were a major concern in many colonial and metropolitan areas—including Mexico City, the capital of the Viceroyalty of New Spain.

CHAPTER TWO

TENTACLES OF COMMERCE

CAMINOS REALES *IN* LATE BOURBON MEXICO

Miguel José de Azanza must have felt miffed. After barely a year as viceroy of New Spain, he found himself in the summer of 1799 responding to an insinuation by royal *valido* Manuel Godoy, the so-titled Prince of the Peace, that Azanza had suspended or altered the important road-construction projects of his predecessor, Miguel de la Grua, the Marquis of Branciforte. Shortly after assuming his duties, Azanza had reported that a labor shortage existed in Orizaba, Xalapa, and Veracruz that affected not only public works, but also that region's entire agricultural sector.[1] The fact that Branciforte was married to Godoy's sister likely meant that he was under attack from his predecessor, which must have nettled Azanza all the more as he assembled a series of reports and documents to append to his letter of explanation. Azanza asserted that as viceroy he had endeavored earnestly to carry out His Majesty's orders, but groundbreaking on the eastern sections of the Mexico City–Veracruz highway had been necessarily delayed in order to accumulate sufficient toll revenues from the already-open section between the capital and the city of Puebla.[2] Spanish colonial rule in Mexico depended in some degree on the ability to communicate and travel between the viceregal capital and the peninsula. Accordingly, the *Camino Real* between Mexico City and the port of Veracruz on the Gulf of Mexico was a continuing concern to colonial and metropolitan governments alike. This chapter analyzes the larger context in which road-construction laborers toiled, an arena of conflict between competing elite interests. The metropolitan government was primarily concerned with the lines of communication between Mexico City and port cities in Spain.

The actions of powerful merchants in the colonial capital, however, demonstrated that their major concern lay with trade and commerce links centered on Mexico City.

THE CAMINO REAL AT MIDCENTURY

The Camino Real, or "King's Highway," served as the arterial connection between Mexico and Spain. Other roads served local markets in one way or another, and the Camino Real did that also, probably much better than it functioned as an artery of the Atlantic economy. Maps prepared at various times during the second half of the eighteenth century indicate the supposed location of the Camino Real's two branches that connected Puebla with Veracruz: one to the east through Xalapa and another to the south, passing through Orizaba and Córdoba.

Travelers' experiences suggest a different reality. In 1766, Fray Francisco de Ajofrín, a Capuchin friar, made an official visit to New Spain at the direction of the Crown in Madrid. The priest traveled from Veracruz to Orizaba, making the journey west from Veracruz on foot. Ajofrín reported that he became lost on the way from Córdoba to Orizaba because of his inability to discern the actual location of the roadbed. He stated that he eventually reached Orizaba and crossed over two rivers via bridges. A 1770 map clearly shows the road and bridges, so one must surmise that Ajofrín was never really very far away from the official path of the Camino Real and that if he could not see it on account of the vegetation cover, then the traffic density must have been very low indeed. In the 1780s, Puebla merchants expressed great concern and interest regarding the proposed routing of an improved *camino real* (literally, "royal highway") between Mexico City and Veracruz.[3] But almost thirty years after the Capuchin friar wandered in search of the road, the situation had changed little. Indeed, an unnamed official in 1794 went so far as to append a note to a letter from Viceroy Juan Vicente de Güemes Pacheco y Padilla, the second Count Revillagigedo, that no road existed at all between Mexico City and Veracruz.[4] Figure 2.1 shows the roads considered in this study; they were not completed until a few years after the viceroy's letter was written.

The colonial government devoted a great deal of concern and effort to the improvement of overland transport during the late eighteenth century, but the problems posed by terrain and distance were immense. The eastern slopes of the Sierra Madre Oriental catch a great deal of rainfall

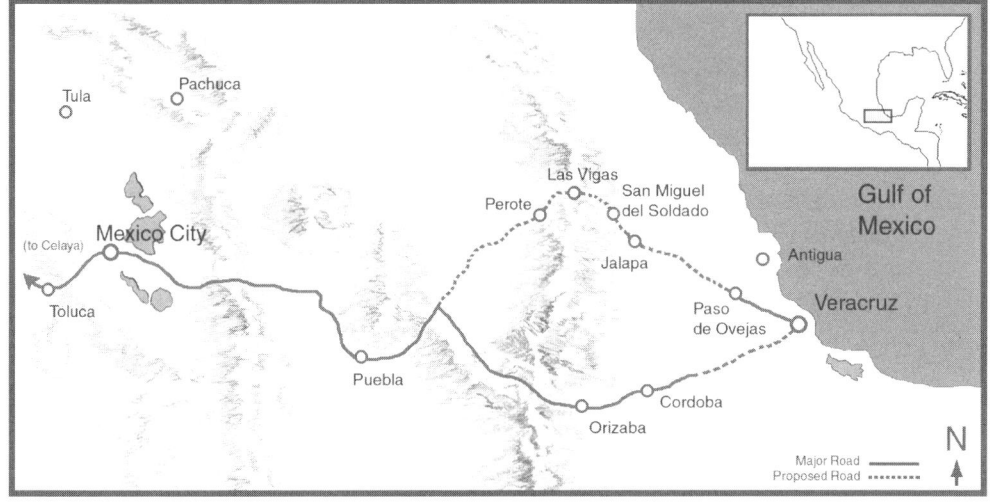

Figure 2.1. Camino Real Network in Mexico, Late Eighteenth Century.
MAP CREATED BY JASON CLARK.

from the northeast trade winds, making the first 150 kilometers west of Veracruz a climb through wet and thick vegetation. Keeping those easternmost legs of the Camino Real passable represented a well-nigh impossible task, even for a Hercules in strength and a Sisyphus in tenacity.

The pestilential conditions in tropical, lowland Veracruz dictated that the Spanish trade fairs be held in a healthier clime, such as that of the pueblo of Xalapa de la Feria. In 1757, for example, we find considerable attention to the roads given by Xalapa's *alcalde mayor*.[5] Records analyzed in detail in chapter 3 show that 2,893.5 *varas* of paved road had been completed in a few key locations, and that work continued on two bridges near Cerro Gordo.[6] Converting this figure to kilometers using thirty-three-inch varas, we find that 3.55 kilometers of pavement were built. A decade later, another alcalde mayor of Xalapa reported completion of 2,551 varas, or 3.12 kilometers, around the Totonac pueblo of San Miguel del Soldado, located to the west of Xalapa along the Camino Real.[7] We shall see later that even these small achievements represented a significant strain on available resources. The Camino Real obviously had a long way to go to become a ribbon of pavement between the capital and the port city. The expense of such a project would have been horrific, and it seems very likely that all that could be reasonably afforded were improvements at sites of importance to local markets as well as at strategic bridges and mountain passes.

The projects around Xalapa appear to have amounted to a Sisyphean effort of laboring to maintain the status quo, but at least one time the roadbed's path was altered in an apparent attempt to improve its quality and to reduce the burden of its upkeep. In 1781, a decision was taken to modify the route of the Camino Real and run it through a maize field that was the communal property of the indigenous villagers of San Miguel del Soldado. Their alcalde protested vigorously. In his letter to Viceroy Martín de Mayorga, don Simón Nicolás protested that his villagers had owned that land "since eternity." If it were taken for the Camino Real, the villagers would lose an important source of their food, of their tribute to the government, and of their material support of the church.[8]

The Audiencia de México (a high court) looked into the matter, and the *oidor fiscal*[9] received a number of pleadings, each attesting to the dire impact that rerouting the Camino Real would have on the villagers of San Miguel del Soldado. Their protestations came to naught, for the *audiencia* decided against the villagers and denied their alcalde's petition. The oidor fiscal acknowledged the seriousness of San Miguel del Soldado's plight, but on August 6, 1781, he declared that the interests of the state in facilitating agriculture, industry, and the military defense of the viceroyalty had to take precedence.[10]

Five months may appear like a rather long time to render a decision that would seem to have been a foregone conclusion. Perhaps the delay until August was intended to allow the villagers to harvest one last crop of maize from the field while the matter was before the audiencia, but that seems unlikely because maize harvests normally began in November.[11] Perhaps the pueblo indeed really did have a fighting chance of success. Many historians have noted the many successes enjoyed by indigenous villages that used the Spanish colonial court system to protect their interests, but such was not to be the outcome in San Miguel del Soldado in 1781.[12] The Camino Real already ran through San Miguel del Soldado before that pueblo litigated the condemnation of their maize field to reroute the roadbed. Their loss of the case did not, in all likelihood, significantly degrade the villagers' environment any more than it had been degraded for the many years in which *arrieros* (muleteers) and their mule trains had tramped through the area. Loss of the field itself still must have hit the pueblo rather hard, as it must have reduced their ability to produce foodstuffs for their own consumption, to sell them in the marketplace, and to use them in meeting their tribute requirements.

Maintenance of the roadbed over the difficult terrain between Veracruz and Perote proved an expensive and continuing thorn in the colonial regime's side, a problem exacerbated by the wet weather and considerably strained government finances. A Spanish military engineer, Captain don Miguel del Corral, surveyed the situation in early 1784. Corral recommended to Viceroy Matías de Gálvez that the burden of labor be placed on all of the indigenous villages along the route. Possibly taking account of harvest labor requirements, he recommended that in November of each year a team of engineers, accountants, and indigenous laborers inspect the condition of the roadbed and effect all necessary repairs. Apparently intending that the primary beneficiaries of the repairs bear the bulk of the cost, Corral recommended that these repairs be funded from local toll receipts.[13] Viceroy Gálvez approved Corral's recommendation and ordered implementation in November 1784. Possibly to ensure the continued presence of the draft-labor source, the viceroy also ordered that militia be used to oversee the efforts.[14] Archival records from the 1780s documenting road repairs around Xalapa apparently do not survive, so the actual extent of implementation of Corral's proposal remains unclear.

VICEREGAL PLANS FOR IMPROVEMENT OF HIGHWAYS

The colonial government's interest in improving the condition and passability of the Camino Real was an ongoing concern, but one that appears to have failed to translate into effective action. In looking at the employment records from a labor standpoint, one also sees that a considerable expense of human and material resources was required every year to work on the road between Xalapa and Perote. Year after year the work sites remained the same, as apparently did the work that was required to be done. The efforts around Xalapa appear to have been attempts to stave off further deterioration rather than make improvements on some baseline condition.

Under Viceroy Carlos Francisco de Croix, a plan for improvements emerged in 1768. The need for viable roads was acknowledged, especially one connecting the capital with Toluca, the breadbasket to its west. All road construction, however, had to wait until fortifications could be built.[15] Veracruz had temporarily fallen to pirates in 1683.[16] A more passable road down to Veracruz might indeed have improved commerce with the Caribbean and with Spain, but without a strong permanent garrison at each of the mountain passes such a road could also have constituted a liability that exposed Mexico City and silver-mining centers such as Guanajuato and Zacatecas to capture by a British enemy. The threat of war with England in 1774 appears to have led to a review of the plan. Christon Archer discusses the subsequent military staff work in his analysis of the strategic importance of Veracruz.[17] A committee of military officers was established to identify problems and to make recommendations for the defense of the Kingdom of New Spain. The panel naturally included several military engineers, officers who were expert in the construction of roads, bridges, fortifications, and the like. As one would expect, the roads from Mexico City to Veracruz served as the subject of a large part of the final report that the officers submitted on January 10, 1775, to Viceroy Antonio María de Bucareli.[18]

The committee's concerns lay principally with military matters. Their assessment of the physical difficulties impeding the movement of either Spanish or enemy columns in either direction along the Camino Real in eastern New Spain, however, applied equally well to mule trains laden with commercial goods. Thus, the army and the muleteers shared many of the same concerns about problems with the kingdom's roads. Indeed, had a blue ribbon panel of mule drivers been assembled to report on the state of highway commerce, they would probably have come up with a document much like the army's recommendations for the defense of Mexico.

The committee stated that although any intelligent person could see the virtual impossibility of preventing the landing of a seaborne invasion force at Veracruz, the attackers would remain unable to move inland until they had captured the Veracruz harbor fortress of San Juan de Ulúa. The well-known problem of disease made Veracruz a poor choice for a permanent cantonment, so the committee preferred garrisons at Xalapa and Perote on the north road, and at Orizaba and Córdoba on the southern branch. These forces would advance along the roads to Veracruz and relieve the expected siege of San Juan de Ulúa. The marching times from the places most remote from Veracruz—namely, Perote and Orizaba— were estimated at approximately five days.

The road toward Xalapa started out with a sandy surface and a gentle slope, but west of Cerro Gordo the climb became steep and difficult. The section from Xalapa to Perote was even worse, especially that part running from San Miguel del Soldado through the mountain pass to Las Vigas, and therefore the terrain offered many advantages of defense. If an invading force did capture San Juan de Ulúa, its next task would be to move inland, which would require the building of provisional bridges at several locations. Obviously, then, several points along the road were river crossings without bridges. One can easily infer that the transit of an arriero's mule train along the Camino Real was impeded by these river fords, and one is then forced to conclude that amelioration of such impediments to land commerce with Veracruz did not enjoy a sufficiently high standing in the viceroys' concerns to warrant the expenditure of a substantial quantity of resources.

Similar difficulties abounded along the Camino Real from Veracruz to Orizaba, although the terrain did not become especially difficult until the approaches to Córdoba. Around Córdoba and along the way to Orizaba lay many hills and ravines, including the Metlác ravine, a very deep canyon at Fortín de las Flores, which at least had a rickety wooden bridge across the river at its bottom. The committee opined that an invading force might possibly overcome the terrain obstacles and get as far as Maltrata, an *yndio* village immediately to the west of Orizaba. They would, however, have found the going very difficult indeed.

The committee officers emphasized that the weather would make it very difficult for an invader to move artillery along the muddy roads from Veracruz to the highlands, which obviously implies that a Spanish move down from the mountain garrisons would be almost as tough. The plan's authors noted that improved road surfaces would speed the relief columns on their way, but they seem not to have argued the point as forcefully as they might have done. They did urge caution in depending on difficult terrain as a major factor in defending New Spain from invasion. Reading between the lines, one suspects that the officers may actually have intended to point out, in a backhanded and politically acceptable way, the advantages that the terrain and unimproved road actually afforded the Spaniards in protecting Mexico City and the silver mines from capture. The defense of the Kingdom of New Spain may well have dictated keeping the Camino Real in a primitive state, and, as shown later, another twenty-five years would pass before a serious but costly attempt at constructing a good highway to Veracruz would be undertaken.

All of these difficulties in moving troops would, of course, also impede the flow of commerce and undoubtedly did exactly that. The defense plan contains a reminder that although wagons had been used to supply Spanish troops during the recent war with Portugal, such an expedient was clearly impossible in eastern New Spain. The backs of draft animals and litters drawn by them were the only viable means of transport available. Wagons and carts could not be used at all, and as one historian has noted, commercial traffic was effectively limited to mule trains.[19]

The defense plan also reveals the existence of two additional roads eventually leading to Mexico City. Located between the two branches of the Camino Real and running more or less parallel to them, these roads rejoined the Camino Real between Perote and Puebla after passing through the mountains and around the Cofre de Perote. The committee members observed that these two roads were in even more lamentable condition than the Camino Real and offered but little access to water. One must infer that arrieros and their mule trains would have found that the costs of delays from these arduous and time-consuming passages around the *garitas* (toll gates) negated some of the attractions of avoiding payment of the tolls.

In all probability, mule-borne freight must have been confined largely to the two branches of the Camino Real between Veracruz and the central plateau. The roads were tough enough obstacles in themselves. Consider the difficulties experienced in outfitting the fortress built at Perote. Its cannons had arrived at Veracruz in 1779, but they weighed so much that officials feared to transport them on the Camino Real via Xalapa. Viceroy Martín de Mayorga appropriated monies from the *alcabala,* or sales tax, so that whatever work was needed could be done to get the cannons over the mountains.[20] Four years later, however, the unexpended funds were returned to their source.[21]

The record does not clearly show how many years passed before the artillery pieces actually made it to Perote. Presumably there occurred some bureaucratic delay in realizing that the leftover funds should be returned, but one suspects that the guns spent a long time in Veracruz awaiting shipment. The Perote fortress played a crucial role as a second line of defense in the 1775 plans, so movement of the king's cannons surely rated high on the list of official concerns, particularly during a period of war with England.[22] That so much time was required to move an important shipment of heavy armament up the road to Perote underscores the difficulty of shipping large quantities of commercial goods along the same route. Perhaps even more significantly, the return of unexpended highway repair funds that had been appropriated years earlier suggests that the viceroy and other influential persons in Mexico City did not actually consider improvement of the Camino Real near Veracruz to be a major concern. As we have already seen, there was always much work to do, so passing up an opportunity to have some of it done speaks volumes about the true priorities in Mexico City. Those priorities would somewhat change after another decade passed, but in 1783 lip service was the order of the day.

THE MEXICO CITY–TOLUCA ROAD

The devastating famines of the mid-1780s resuscitated concerns about the Mexican transportation infrastructure. Crop failures in many central Mexican locales led to widespread starvation, a disaster of such magnitude that population totals visibly declined. The colonial government attempted to ameliorate the suffering through its reactions at various levels of administration, but they could do little, and they did it too late.[23] A more viable solution seemed to be promised by construction of a paved road between Mexico City and Toluca, a suggestion that by then had been on the table for more than twenty years.[24] Toluca had extensive commercial ties to Mexico City throughout the last half of the eighteenth century.[25]

Colonel don Bernardino Bonavía, intendant of Mexico and *corregidor* of Mexico City, made a concrete proposal to Viceroy Revillagigedo in 1791. The "very important" road to Toluca should begin forthwith, to be funded by toll collection and by money to be borrowed from the residents of both cities and the *hacendados* in between them. Bonavía signed out a letter charging Captain don Manuel Agustín Mascaró, a military engineer, with reconnoitering and mapping the optimum route for the road.[26] Mascaró's sketches and maps were highly regarded, as one can see for oneself in figure 2.2. He set out immediately for the countryside and occasionally submitted interim reports. From Lerma in May 1791, he reported that the wheat crop in the Toluca region was quite bountiful, listed the various pueblos and haciendas he had visited, and acknowledged the hospitality of various important people.[27] A month later he gave an initial estimate of 102,331 pesos to construct the road, commenting that a highway of such length (sixty-four kilometers) was a huge project that would require many structures to keep it in operation once it was opened. He went on not only to reemphasize the extreme value that the planned road would have for Mexico City and the valley of Toluca, but also to opine that the benefits would ripple all the way to Nueva Galicia and foment the development of industries such as textiles.[28]

Without money, of course, nothing tangible could happen. Viceroy Revillagigedo and the Consulado de México, a guild of important merchants in Mexico City, offered financial support in the form of loans from many of the merchants and landowners who stood to benefit from the new highway. As one might expect, the willingness to participate varied from contributor to contributor.[29] "Contributions" were also exacted from those apparently expected to make the most direct use of the new road. Captain Mascaró had also been placed in overall charge of toll collection, which began immediately all along the route in the spring of 1791, almost two years before the first paving stone was actually laid! In the matter of toll collection, Mascaró was assisted by don José Sanz, the intendant of the Real Hacienda. In May 1791, Sanz reported that he had encountered significant resistance to toll collection and that he had hired several officials to administer and enforce the new toll schedule.[30]

Mascaró continued with his engineering work, but in a letter written a few weeks after Sanz's he reported that the crafty arrieros used many tricks to avoid toll payment. Enforcement costs ran 2,300 pesos. Approximately 15,008 pesos had been collected thus far, leaving 13,700 for the construction fund.[31] One can easily sympathize with the arrieros' resistance to the implementation of a toll collection scheme along a series of unimproved pathways and roads that they had for years used without charge, especially considering that the improvements would be a long time in the making.

In the meantime, an alternative plan was developed by don Felipe Narvarte, but Mascaró's earlier plan was chosen, and the latter was placed in overall charge of construction.[32] After the passage of almost *another* year, Viceroy Revillagigedo reported to His Majesty that the construction fund had finally amassed enough money to proceed. The long-awaited Mexico City–Toluca highway project was to begin immediately, notwithstanding the continued objections of certain landowners, who themselves would

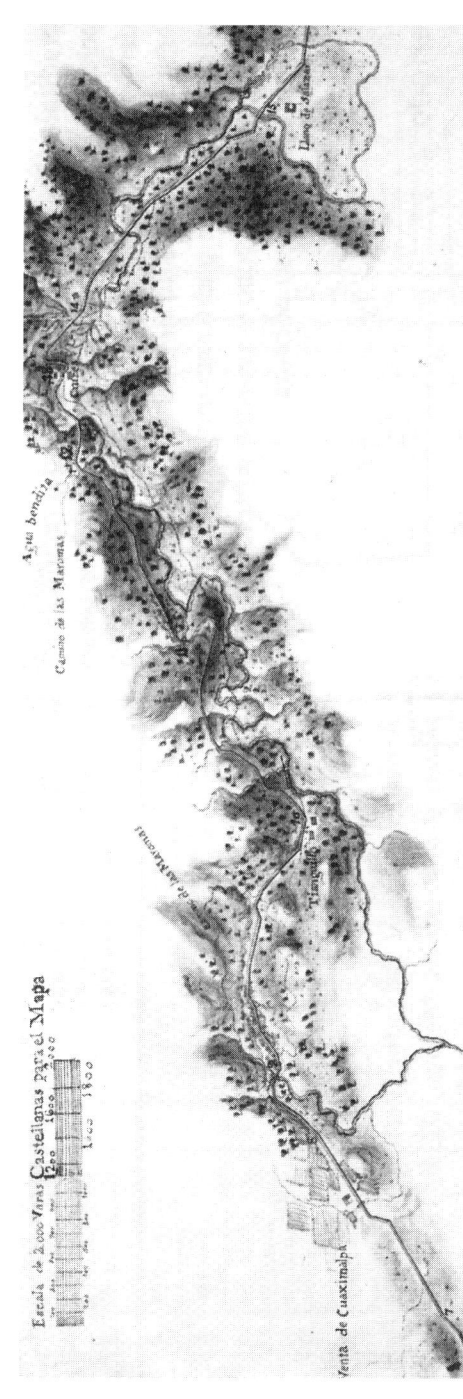

Figure 2.2. Mexico City-Toluca Road Segment. Map by Captain don Manuel Agustín Mascaró, 1791. AGN Caminos y Calzadas, Vol. 11, f. 284 (7). Courtesy of the Archivo General de la Nación, Mexico City.

ultimately benefit—regalist absolutism at its best![33] The road surface was to be built in eight sections.[34] Two groups, the Brigada del Poniente under Captain Mascaró, and the Brigada del Levante under Captain don Diego García Conde, worked simultaneously on separate sections of the roadbed. Construction began on November 14, 1793, at a point somewhere between the two cities, with the Brigada del Levante working eastward to Mexico City and the Brigada del Poniente working westward to Toluca.[35]

Work proceeded for about two years until the two construction brigades reached their respective destinations. A shortage of funds in 1794 led to cost-cutting measures, including wage reductions, which are considered in chapter 4. Mascaró appears to have requested to be relieved and posted to Spain, but a reply from Madrid noted the extreme importance of the Toluca road, applauded the engineer's diligent work, and directed that he remain in charge of the project.[36] With the completion of the highway to Toluca, Mexico City would acquire a strengthened tentacle into an expanded hinterland.

Whether the methods used by the colonial regime to finance construction of this major highway are unique to "colonialism" is a question that requires comparing methods used in other "colonial" or "noncolonial" situations. In the late eighteenth century, expansion and improvement of the transportation network was a major concern in the southern Spanish region of Andalucía. *Repartimientos* levied against local governments were a major source of such funds. These particular repartimientos assessed money as opposed to drafting persons from the jurisdictions, so the recruitment source of the actual laborers remains obscure. Landowning Andalucian nobles enjoyed considerable success in resisting levies by the Council of Castile against themselves, a success not enjoyed by their hacendado counterparts between Mexico City and Toluca. In Andalucía, the responsibilities for action on public works were assigned to the intendants who governed the various areas.[37] In the case of the Toluca highway, the viceroy of New Spain himself exercised considerable direct authority. The Consulado de México was also actively involved. Control in this colonial situation was exercised at a higher level, with the direct involvement of both Viceroy Revillagigedo and Intendant (and Corregidor) Bonavía. Chapter 4 deals with construction of a stone highway bridge in the Intendancy of Veracruz. In that case, responsibility was assigned to the Ayuntamiento de Orizaba, a municipal entity at a lower echelon than a provincial intendancy. Control and authority in both Spain and Mexico thus rested with a local authority, but in the cases under consideration here, they used different methods to raise the funds.

THE CAMINO REAL TO VERACRUZ

The Consulado de México proposed expansion of the highways of New Spain far beyond the significant but beginning step represented by the completion of the road between Mexico City and Toluca. This merchants' guild played a crucial role in developing commercial networks centered on the capital, of course feathering their own nests all the while.[38] Their focus shifted toward the Camino Real to Veracruz. The conditions described in the defense plan of 1775 appear to have changed but little in the region lying east of Perote and Orizaba. Recall that an official in Madrid had declared that no road existed at all between Veracruz and Mexico City. The Marquis of Branciforte succeeded Count Revillagigedo as viceroy of New Spain in that year and then directed considerable personal attention to the highway network.

Perhaps earlier viceroys had indeed read between the lines of the 1775 defense report, taken comfort in the military advantages to defense against invaders offered by the difficult terrain east of Perote and Orizaba, and then chosen not to undertake major projects to improve the roadway to Veracruz. Considering the hampering effects that state penury had on the Bourbon reforms, however, one might contend with equal plausibility that the viceregal regime simply could not afford the public works, so that other arguments became moot. By 1793, however, a desire to enhance commerce with an improved network of roads appears to have become an uppermost concern of the viceroy and other important personages in Mexico City. We have seen the relative speed with which the Mexico City–Toluca road was constructed during 1793–95. This project enjoyed great support from both viceroys, as seen in Revillagigedo's reasons for dismissing local landowners' complaints when Captain Mascaró made last-minute changes to the road's intended path. Revillagigedo brushed aside the petitioners and pointed out the extreme importance of that particular road to the entire kingdom of New Spain.[39] The influential merchants of the Consulado de México stood to profit handsomely from anything that expanded commerce, an interest that dovetailed nicely with some government and church officials' hope that the ability to move grain more quickly might lessen the impact of future crop failures like the disasters of 1785 and 1786.

Intense competition erupted between the Consulado de México and the Consulado de Veracruz over the route to be used from Puebla to Veracruz. The *real cédula* formally establishing the latter institution would not be received until 1795. In effect, however, the merchants who would become its members had long acted as a de facto interest group since at least 1781, when they first petitioned the Crown for incorporation as the Consulado de Veracruz. In particular, the Veracruzano merchants benefited from Viceroy Revillagigedo's favorable disposition toward themselves and their ideas.[40] The controversy burned for years. The Veracruzanos preferred the northern option via Xalapa and Perote and contended that the northern route was shorter than a southern one through Orizaba, a contention that any atlas shows to have been simply not true. (See figure 2.1.) They also argued that the slope was easier along the northern route and also that construction materials and labor were more readily available. They emphasized the importance of good roads to great states because such roads facilitated both commercial traffic and military movements. They went on to highlight the possibilities for development of a commercial agriculture sector around Veracruz and pointed out that a good road that led west from the port city would mean that crops could be distributed in Mexico as well as exported by sea.[41]

By equating commercial expansion with increased state power, the Veracruzano merchants echoed themes that figured prominently in the Bourbon period. Looking behind their emphasis on the benefits to the entire kingdom of a road from Veracruz to Mexico City through Xalapa, we quite easily see that these same persons stood to profit greatly with the expansion of commercial agriculture in the area immediately adjoining their city. Consulado de Veracruz merchants maintained extensive commercial and residential interests and connections in Xalapa and Perote.[42] Xalapa's importance to Veracruz merchants also stemmed from its role as the site of the trade fairs for Spanish goods.[43]

The Consulado de México found an ally in Viceroy Branciforte, who supported their desire that a paved highway from the capital to the coast run through Orizaba and Córdoba. Any viceroy of Mexico was a very important person in the Spanish Empire, but recall that Branciforte had a particularly valuable connection as the brother-in-law and protégé of influential royal advisor Manuel Godoy.[44] In a 1796 letter to Godoy at Aranjuez, Branciforte vented his frustration with the Consulado de Veracruz and its continuing opposition to his plans. The viceroy declared that the Xalapa route would benefit only Xalapa itself, as it was not the shortest route. So annoyed was Branciforte that he recommended that the king disestablish the Consulado de Veracruz and the Consulado de Guadalajara, likening them to small monsters that devour each other. He emphasized his continuing interest in the Orizaba route, which would facilitate the more rapid movement of tobacco to Mexico City and Veracruz. Branciforte pointed out that he lacked the funds to undertake a worthy project of such great magnitude and that using the toll funds collected on the Mexico City–Toluca road would remove the revenue to a place far from where it was paid.[45] Godoy made no written comment regarding the consulados, but he did take action on the question of highway construction.[46]

Only one day after Branciforte fulminated on official paper about the obstreperous consulados, he sent yet another letter to Godoy requesting monetary support for the construction of the roadway to Veracruz through Orizaba. He declared that it was the most important issue then facing him as viceroy. Even with 26,000 pesos of annual profit from the Toluca tolls, the government of New Spain still found itself without enough capital to proceed. Everyone in Mexico was doing all he could, such as the Count of Contramina, who surrendered all of his income from the Toluca tolls and added his own funds as well.[47] Contramina's own funds were presumably added through his purchase of government bonds instead of through outright donations.

Led by the Count of Contramina and wealthy merchant don Antonio Bassoco, the Consulado de México within one week went on to advocate yet another major road-construction project—extending the recently completed Toluca road to Celaya. The consulado had already built five bridges toward Ixtlahuacan, one of which was quite large. Such a road would allow wagon and road transport between the Bajío and Mexico City. Ultimately, the consulado suggested, Veracruz could be connected with the Pacific port of San Blas, which would reduce the costs of transport in the kingdom and (its members claimed) facilitate the more rapid movement of silver bullion. Bassoco and Contramina recommended procuring royal funds for the Celaya project and urged the expeditious commencement of the new road to Veracruz, especially the section between Mexico City and Puebla.[48]

The response from those primarily interested in the Orizaba-Veracruz road came posthaste. The intendant of Puebla, don Manuel de Flon, urged that construction begin immediately on the project because it would greatly benefit the entire kingdom. Don Marcos González and other members of the Ayuntamiento de Orizaba pointed out the importance of road improvements to alleviating food shortages in remote villages and spoke of the local arrieros' constant clamor for those improvements. They recommended that the proceeds from their toll collection at Orizaba be used to build this road.[49]

Smaller places also pleaded their interests to Viceroy Branciforte concerning selection of the route to be followed by the new Veracruz road. Including both the governor of the *españoles* (meaning here everyone who was not legally an yndio) and the governor of the yndios, eighteen villagers from Ameca proposed that the section between Mexico City and Puebla come their way instead of through Río Frío. They argued that construction costs would be reduced because fewer bridges would need to be built and pointed out that a large supply of labor would be available to build the road in the area of their village. In addition to claiming that Río Frío was a place with but a small population, the Ameca villagers also attempted to convince the viceroy that construction would suffer unnecessary delays if the Río Frío route were chosen.[50] Their blandishments proved to be of no avail, and Río Frío remained notorious as a beehive of banditry on the Camino Real.[51]

Branciforte seems to have been at least temporarily swayed by Contramina and Bassoco's enthusiasm for the Celaya road extension. In his letter of October 27, 1796, the viceroy remarked to Godoy on the eagerness displayed by Puebla and Orizaba for the proposed eastern highway construction, but went on to state that the Celaya road was undoubtedly the most valuable project before him at that time.[52] The last remark is surprising, but that particular letter contains the only indication that Branciforte ever gave priority to any proposal other than the new road to Orizaba and Veracruz, once the Toluca highway had been completed.

Branciforte's subsequent actions indicate that if he had ever temporarily harbored doubts as to the primacy of the Veracruz road, they had since been washed away. On December 9, 1796, work began on the new road to Veracruz. To commemorate the groundbreaking, Branciforte commissioned sculptor Manuel Tolsá to begin a bronze equestrian statue of King Charles IV in the Plaza Mayor. Well known today in Mexico City as "El Caballito," the statue was completed in 1803.[53] The road was named "Camino de Luisa" in honor of Queen María Luisa, on whose birthday the work began.[54]

Viceroy Branciforte did not submit a report that construction had commenced until December 29, 1796.[55] Incredibly, he had not yet received authorization from his superiors in Spain to begin the work. The impatient Branciforte had probably committed to begin on the queen's birthday and found himself frustrated when approval of his September recommendations did not arrive in time. On the day preceding his announcement to Godoy, the viceroy sent him still another letter commenting on the unusually long interval of time that had passed since he had last received mail from the palace at Aranjuez.[56]

Two months after the groundbreaking ceremony, one Pedro de Mantilla wrote to Godoy from Madrid, contending that Branciforte had been deceived by the Consulado de México. The latter did not wish to share power with any other organization in New Spain and had managed to preclude the Consulado de Veracruz from being heard. Appointed by the Consulado de Veracruz to represent its interests at the royal court,[57] Mantilla went on to argue the merits of the Veracruzano case for the Xalapa route, claiming it had better terrain because there was only one major obstacle, the Río Antigua, which required a lengthy bridge.[58] The Consulado de Veracruz thus appears to have attempted to circumvent Viceroy Branciforte and to deal directly with Aranjuez on a matter of great financial importance to its members.

The Crown decided in favor of Branciforte and the Consulado de México, issuing a royal order on February 11, 1797, more than two months after the work had actually started. Both the extension of the Toluca road to Celaya and a new highway from Mexico City to Veracruz via Orizaba and Córdoba received approval. Use of toll receipts from the Toluca road was also authorized to pay for each of the two roads.[59]

What then to make of Viceroy Branciforte's having commenced the Veracruz road several months before he could possibly have received the document containing the king's authorization? He possibly had received word through an unofficial, or at least unrecorded, channel that his case had prevailed at Aranjuez, but that seems extremely unlikely in view of the short time intervals between the documents that do survive in the archives. The viceroy may have felt certain of his brother-in-law's support, or perhaps he meant to force the Crown's hand by presenting Godoy with a fait accompli, believing that Aranjuez would never act to undermine his viceregal authority. The December 29 letter conceivably *could* have reached Spain by February 11, but the reaction in Aranjuez would have to have been much swifter than was usually the case with letters from Mexico City. Most likely, Godoy had not received Branciforte's December letters when he issued the approval order in February, and the viceroy had simply gambled that he would prevail. This may seem a rather impetuous move by Branciforte, but one must recall that this same man not only wrote a letter venting his pique and recommending abolition of the Consulado de Veracruz, but actually posted it to Spain, an imprudence that other persons might have avoided after a cooling-off. References to the Toluca-Celaya road cease in archived correspondence from this point onward, but construction of the first leg of the Camino de Luisa appears to have proceeded as a matter of high priority.

Eighteen months later the Consulado de México proudly reported to the new viceroy, don Miguel José de Azanza, that the first third of the new highway to Veracruz had been completed. All classes of passengers, coaches, and wagons could now travel from Mexico City to Puebla. The Puebla road was a spectacular achievement. It had a stone-paved surface, and trees had been planted all along the roadway. The consulado boasted of its new bridge at Chalco, raising to four the number of such spans along the road to Puebla.[60] What was intended, at least on the official record, to have been a major artery of commerce to Veracruz at this point stopped well short of its stated distant goal.

Orizaba and Córdoba in the meantime continued their efforts to improve the Camino Real in their vicinity. The two *villas* found themselves in 1797 embroiled in a dispute over local priorities. Twenty-one prominent Orizabeños, several residents of other pueblos as far away as Tepeaca, and even some arrieros petitioned the viceroy for funds to improve the roads leading west to Maltrata and Acultzingo. They complained that unusually heavy rains had reduced road traffic, and consequently only 4,329 pesos were available for the important and necessary work. By

August of a normal year, they would have collected at least 6,000 pesos above expenses.[61]

The Ayuntamiento de Córdoba reminded all concerned that their villa's residents paid a great deal of the Orizaba toll intake. They protested that the road running east from Orizaba to Córdoba should be repaired before taking care of such small places as Maltrata and Acultzingo. With this repair, it would become much easier to move tobacco leaf from Córdoba fields to the royal factory at Orizaba. The dispute fell to the intendant of Veracruz for resolution. He released some money to the Ayuntamiento de Orizaba, but instructed that it be used to repair the Camino Real between there and Córdoba.[62]

Those who lived along the proposed second construction leg of the Camino Real to Veracruz clearly did not wish merely to wait for the Consulado de México's project to reach them. The Ayuntamiento de Orizaba obviously intended to extend its own tentacles and expand its hinterland to include Maltrata and Acultzingo more effectively. By overlooking Córdoba's political power and the importance of its role in the local market, the Orizabeños invited the unspoken rebuke that the intendant's decision represented. But although the decision involved the Camino Real, it affected only the configuration of the local market and addressed no issue regarding trade links with Mexico City or with Spain. Moving tobacco from Córdoba and Orizaba to Veracruz for shipment overseas remained quite difficult because nothing had been done to improve that section of the Camino de Luisa.

This is the point, discussed at the beginning of the chapter, at which Viceroy Azanza appears to have been taken to task by Godoy. The Consulado de México reported to Azanza that it had intended to carry out the royal order of February 1797 by constructing the new highway in three parts. The first ran from Mexico City to Puebla via the Río Frío mountains (not through the pueblo of Ameca). The consulado had delayed beginning the second leg until the Mexico City–Puebla road had been opened long enough to accumulate sufficient toll receipts to bankroll additional construction. The merchants expected to complete the next leg in a timely fashion, but they complained of a shortage of labor for such a project. Road construction was always slowed in New Spain, they claimed, because harvesting crops necessarily exercised a higher-priority demand on the labor force.[63]

Azanza assessed the Mexico City–Puebla road as useful, even necessary, to the kingdom. Two competent military engineers, Captain don Miguel Costanzó and Captain don Diego García Conde, had surveyed the route east of Puebla to Orizaba and reported that the terrain posed many costly obstacles to construction. The viceroy reported that the Consulado de Veracruz had resumed advocating a new road to Xalapa. Azanza reviewed the proposals and the cost estimates and concluded that the Orizaba route should be retained in the plans. In an apparent sop to the Consulado de Veracruz, the viceroy suggested that a branch road could be built to connect Xalapa with Veracruz.[64]

Diversion of Toluca toll receipts for the construction of the Camino de Luisa appeared to have ceased following the completion of the Mexico City–Puebla section. The Consulado de México's intention that the rest of the road should wait until funds had been accumulated from the Puebla road speaks volumes about the true priorities of the proponents of the Camino de Luisa. Toll receipts could be siphoned out of the Toluca area to build a road to Puebla, but not to Veracruz or even to Orizaba. Merchants of the Consulado de México, such as the Count of Contramina or don Antonio Bassoco, apparently became less willing to invest in the undertaking. One historian argues persuasively that Bassoco in particular possessed a very shrewd business acumen.[65] By 1799, the merchants of Mexico City had much of what they seem to have really wanted: a good road running west to Toluca and another one running east to Puebla, with the capital city at the hub. Extending that network through Celaya to San Blas on the Pacific or to Veracruz on the Gulf of Mexico coast may have remained a worthy goal, but at that particular point in time it was still a vision for which the merchants could afford to wait.

The rival Consulado de Veracruz, however, remained unwilling to wait. Its members recorded their desires in a declaration read at a meeting of their governing committee on January 9, 1800. Reiterating the general benefits to commerce of reducing transport costs through highway improvements, the Veracruzanos lamented the dismal condition of the road between Xalapa and their city. Only on a road from Veracruz could goods from Europe enter the interior of New Spain, and because everybody in the kingdom stood to benefit from a passable Camino Real, the state should fund its construction. They argued that the route through

Xalapa was shorter and over gentler slopes, with better access to building materials, so that their proposed road could be passable along its entire length even to wagons. The knotty problem of crossing the Río Antigua at long last appeared surmountable because Captain don Diego García Conde had located a suitable site for a bridge.[66]

Accusing the Consulado de México of "forgetting" the work that would highlight the wealth and grandeur of both Mexico City and the Spanish Empire, the Consulado de Veracruz proposed that the costs of building the Camino Real be covered through tariffs levied on European goods bound for the interior. Those who benefited most would thus bear the cost burden, as had been the case in funding the Toluca road. The Veracruzanos proposed that places such as Córdoba, Orizaba, and even Puebla could then build branch roads off of the Camino Real, which would move agricultural products to market and thereby extend their respective hinterlands. Those cities, of course, would have to pay for their branch roads by themselves.[67] The Consulado de Veracruz's vision for its city included development of an agricultural cornucopia that would supply all of New Spain.[68] The merchants thus displayed their hopes to diversify their interests and become more than commercial middlemen, but they never retreated from their contention that the state should pay for the highway from which they themselves stood to gain a great deal.

After 1800, the Camino de Luisa appears to have lost all momentum. The arguments coming from the Consulado de Veracruz may have swayed the opinion of Viceroy Azanza or of his successor, don Félix Berenguer de Marquina. The Ayuntamiento de Orizaba reacted in apparent frustration. By 1802, four years had passed since the completion of the Mexico City–Puebla leg of the Camino de Luisa, but nothing had been done to extend it onward toward the Gulf of Mexico. An Orizabeño representative went to Mexico City and approached the Consulado de México on the matter of the road ordered by Branciforte. The person's identity is not revealed by the record, but he appears to have been very outspoken, and he made a clear impression on the consulado merchants. They advised the ayuntamiento that if the viceroy had clearly wanted the work on "Branciforte's road" to go forward, then the consulado would have received clear orders to press ahead. Many in the consulado supported extending the road through Orizaba to Veracruz. The *capitalinos* of Mexico City then chided the Orizabeños with a reminder that noisy people appear rude, hurt the causes for which they speak, and look ill-informed or even ill-intentioned.[69]

The Consulado de México neglected to state the obvious. Had its members perceived that extending the Camino de Luisa beyond Puebla was in their commercial interests, they would themselves have pressed their case to Viceroy Berenguer de Marquina. As had been the case when Toluca tolls were diverted to construct the Puebla road, the powers-that-be could make things happen if they found them beneficial, which seems another way of stating that there needed to be an acceptable probability that the bonds would be somehow paid off. In New Spain, this meant from toll receipts. The Consulado de México did not, however, enjoy omnipotent influence with every viceroy. Matilde Souto Mantecón and Javier Ortíz de la Tabla Ducasse contend that Revillagigedo clearly favored the Consulado de Veracruz and believed that its Mexico City rival functioned only as an obstacle to progress.[70] In another work, Ortíz de la Tabla Ducasse finds that the Consulado de México used all its force to concentrate commercial activity in the capital.[71] The consulado merchants' actions with respect to important arteries of commerce discussed earlier certainly suggest that they at least sought to dominate activity around the kingdom of New Spain from a hub of power in the capital—an achievable goal if they could garner the viceroy's cooperation.

After succeeding to the viceregency in 1803, José de Iturrigaray pressed both consulados to move forward with their branches of the Camino Real. Lack of funds continued to hamper progress in overcoming difficult obstacles of terrain. Attempting to raise more money, the Consulado de México promulgated on February 28, 1804, a revised toll schedule for the Mexico City–Puebla road, but it left in place many of the existing exemptions from payment.[72]

Don Antonio de Bassoco affixed his approval signature to the revised toll regulation, but he immediately wrote a letter to the viceroy on behalf of the Consulado de México, proposing changes that went beyond what the merchants felt they could approve. Bassoco attributed lack of progress on the road projects to lack of funds, which he blamed on low yields resulting from too many toll exemptions. The regulations exempted sheep because shepherds were always poor. Local yndios with burros paid nothing. At Mexico City's Garita de San Martín, they collected nothing from neighboring haciendas. Persons on public service received discounts, as did all military people, mendicant friars, and employees of the Real Hacienda. Because every person on the road benefited from the road, each of those persons should pay the standard toll, Bassoco argued. In addition to abolition of all toll exemptions on the Mexico City–Puebla road, he recommended additional, across-the-board increases for all categories of traffic.[73]

Iturrigaray approved all of Bassoco's recommendations, except that he established the charges for coaches and carriages as two pesos and for mule-drawn wagons as three pesos.[74] The monies collected at the garitas amounted to a tax on users of the Camino Real. In one sense, then, Iturrigaray placed himself in the position of levying a new tax on indigenous peoples because they had previously been exempt if walking or if only driving burros. Many scholars have discussed the role played by the Spanish Crown in affording its subjects of the *república de yndios* a considerable degree of (subjugated) autonomy and protection of at least some of their land rights.[75] Even that shield from creole encroachment, however flimsy it may have been, was lost after Mexican independence in 1820, and one can see in Viceroy Iturrigaray's actions an example of the ways in which pressure from European Spaniards and American-born *creoles* manifested itself to yndios.

Work on the Camino de Luisa progressed very slowly. Three years later, in 1807, a military engineer en route to inspect fortifications reported to Iturrigaray that the section between Orizaba and Córdoba still proved difficult, presumably owing mostly to the deep Metlác Ravine between Orizaba and Córdoba. Even slower progress occurred on the Córdoba-Veracruz section, which had progressed only as far as Atoyaque, a short twenty kilometers east of Córdoba. The inspector reported that the slopes there were particularly treacherous and that, given the bridges that still needed to be built, the budgeted 350,000 pesos would not be enough.[76]

The Consulado de México petitioned the viceroy for yet another round of toll increases, which Iturrigaray promptly granted.[77] Almost equally promptly, arrieros along the Mexico City–Puebla road protested the increases. Some members of the Consulado de México began to despair of ever completing the project, and they attempted to distance themselves from the matter by claiming that even with ten years and 10 million pesos, no paved road could be built through Atoyaque because of the terrain and weather difficulties.[78]

Iturrigaray seemed ready to give up as well. A letter to the Crown reveals the viceroy's pessimism. He reported that efforts on the Camino de Luisa had been underway since Branciforte's time. He felt the project to be a worthy undertaking because Puebla, Orizaba, and Córdoba were cities with large populations and great commercial potential. Added to the difficulties of geography, however, were problems in collecting tolls on all of the kingdom's roads, and enforcement had come to require stationing militia at every garita. High overhead costs thus reduced the quantity of funds available for actual construction. Iturrigaray stated that he had himself looked at the road from Orizaba to Córdoba and Veracruz and concluded that he could find no reason for any confidence in ultimate success.[79]

The Camino de Luisa project appears to have been abandoned shortly thereafter. Work broke off at the pueblo of Chiquihuite, approximately one-third of the distance between Córdoba and Veracruz.[80] That location corresponds roughly to Atoyaque. Travelers' accounts suggest that thirty years later, the road through Xalapa remained the only viable route from Mexico City to Veracruz, but that route was still a difficult one.[81] One historian points to the poor state of the transportation infrastructure as a key barrier to economic growth throughout most of the nineteenth century in Mexico.[82] Major obstacles had to be overcome in order to effect any significant improvement in the situation.

Financial and geographical difficulties plagued the road through Xalapa no less than the Camino de Luisa. In 1807, a road project commenced at Las Ánimas, a place immediately adjacent to Xalapa and now part of the modern-day city. The director used prison labor from the Presidio de Vergara.[83] The road at Las Ánimas was in fact part of the Camino Real, but the work itself probably was undertaken because it improved communication in and around Xalapa. By 1809, only as far as Paso de Ovejas had the Consulado de Veracruz improved the north branch of the Camino Real to the point where it could support carriage traffic from the port.[84] Not until 1811 would the Río Antigua finally be bridged.[85] After 1812, additional construction became impossible owing to the economic disruptions caused by the independence struggle.[86]

CONCLUSIONS

Each of the road projects discussed in this study, born though they were from visions of grand networks of commerce, ended up as part of a more localized market network. Although much grander in scale, Mexico City with its spokes to Puebla and Toluca still constituted a basically local market. The financial resources to do more simply did not exist in New Spain during the late-colonial period.

The strength and relative importance of economic ties between Spain and its American colonies has been a major focus of Latin American historiography for almost three decades. Many historians such as Stanley J. Stein and Barbara H. Stein emphasize the subordination of the American periphery to the European metropolis.[87] This dependency paradigm emphasizes the importance of international markets in explaining the effects of colonialism in Latin America. Dependency theory has come under attack from historians of widely varying viewpoints. Matilde Souto Mantecón focuses on imperial connections, but finds a great deal of agency on the part of local actors where earlier historians discerned teleological outcomes. One of the important criticisms highlights dependency scholars' tendency to discount the importance of regional and local markets, especially during the colonial period.[88] For example, Robert Patch and Eric Van Young find that throughout the colonial period, local and regional markets were of paramount importance in areas as different as Yucatán and Guadalajara, respectively.[89]

These apparently opposing interpretations often appear to use lines of approach that seem more skewed than in direct contradiction. Take it as true, at least for argument's sake, that Spain's interest in its American colonies was what the colonies could do for the peninsula. In the case of Mexico, then, one would argue that Spanish interests centered on the products extracted from the silver mines. But several major studies of Mexican silver mining suggest that the number of persons directly involved with mining, refining, and transporting silver to Veracruz for shipment across the Atlantic was but a tiny fraction of the population of colonial New Spain.[90] The same can be said of other exports such as tobacco and cochineal. Such economic activities could not be carried out without a local economy to grow food and to create other local products, both of which required a large social and economic infrastructure that would in size far outstrip the mining, tobacco, and cochineal sectors themselves. In other words, large local market economies served as foundations that underpinned comparatively small export structures, and most people's lives were wrapped up in the former category. Historian Richard L. Garner finds that one hundred miles was the maximum radius in which low-priced agricultural goods could be cost-effectively transported, contradicting an earlier study of transportation in Mexico in which geographer Peter W. Rees argues that the Camino Real between Mexico and Veracruz bound the Viceroyalty of New Spain in a dependent relationship with its Iberian master.[91]

The results of this study underscore the continuing importance of local and regional markets, as we have seen in the successful efforts of the viceregal government and the Consulado de México to expand Mexico City's sphere of influence by building highways to Toluca and Puebla. Those cities became enmeshed in a market web centered on the capital. The eastern link later reached all the way to Córdoba, but the final link to Veracruz and to metropolitan Spain remained unimproved, so Veracruz remained difficult to reach by road from Mexico City at the outbreak of the Mexican independence struggle.

CHAPTER THREE

From Draft to Free-Wage Labor

Projects around Xalapa during the Mid-Eighteenth Century

The alcalde mayor of Xalapa de la Feria seethed with frustration. Work on two important bridges near Cerro Gordo progressed but slowly, and these structures would have greatly eased the difficulty of travel between Xalapa and the port of Veracruz. As Xalapa was the site of the (more or less) annual Spanish trade fairs, the Cerro Gordo bridges were no trifling matter. Alcalde Mayor don Antonio Primo de Rivera gave vent to his aggravation in a letter to Viceroy don Agustín Ahumada y Villalón in the summer of 1758. Primo de Rivera projected completing the respective bridges in July and August of that year, but added that they would have been finished sooner but for "Indian reluctance."[1] He attempted to impress on Viceroy Ahumada the extreme inefficiency that he felt stemmed directly from the draft-labor regime used to bring Totonac *peónes* to work on public projects.

The villagers of the various pueblos probably saw little value to themselves in the Camino Real, if in fact they were dragging their feet. That such villages could take action on things that really mattered to them is as much a part of the historical record as officials' complaints about their supposed laggardness. Consider, for example, the pueblo of Coatepec, this one located near Mexico City as distinct from the one with the same name near Xalapa. In 1736, don Diego Pacheco, the yndio governor of the first Coatepec, petitioned the alcalde mayor don Juan de Sayanez for permission to rebuild a road bridge over which the villagers passed on their way to the milpas where they grew their maize crops. This wooden bridge was in such a state of decay that the villagers feared to cross it and believed that in any event the structure would not survive the next rainy season. Pacheco asked to replace it with a stone bridge, to be constructed under the supervision of an "intelligent person" to be provided by the viceroy. The Coatepec villagers offered to bring the stone to the site and provide all the labor themselves. Sayanez rode out to look at the site himself. He concluded that the villagers' fears were quite well founded, and because he lacked the authority himself, he requested that Viceroy don Juan Antonio Vizarrón y Eguiarreta allow them to construct a stone bridge.[2] The record does not tell us what action Vizarrón took, but it seems very likely that he would have approved the Coatepec villagers' request. In essence, the Nahua of Coatepec drafted themselves and used the colonial regime's institutions for their own ends.

In general, infrastructure repairs were the colonial government's concern, and the situation around Xalapa seems to have been more common. The impetus stemmed from well-placed people of influence more often than it did from the indigenous rural villagers who would actually do the work. This chapter analyzes changes in the Spanish colonial regime's methods of mobilizing labor to build and maintain parts of the Camino Real during the eighteenth century.

Draft Road Laborers in Xalapa

Returning to the words of Alcalde Mayor Primo de Rivera and assuming that they can be taken at face value, we can see that an opposite situation developed around 1757 in the pueblos around Xalapa. Although it *may* have been of little practical value to the Totonac of the outlying pueblos or even to Xalapa's own indigenous "Barrio de Jalapa," work on the Camino Real remained important to the colonial regime. Highway improvement required the colonial regime to mobilize manpower, tools, and construction materials. Officials of local governments, such as Primo de Rivera, often exercised direct control and oversight of projects of this sort. Of course, they had no departments of public works, but the political structure of the república de yndios perfectly fit the needs of the colonial regime. The viceroy could simply authorize an alcalde mayor to direct the governors of native pueblos to provide specified numbers of villagers to the *sobrestantes* (foremen) at particular worksites. Although the beginning of a shift toward free-wage labor lay no more than a decade in the future, in

1757 the Totonac pueblos around Xalapa still provided drafts of laborers for road work.

The villages provided draft laborers for road repairs along the so-called Camino Real under the tributary regime described earlier. The men served as day laborers at construction sites located many kilometers from their farms and villages. For this work, each village governor received a single real for each *peonada,* or man-day of work performed.[3] For the year 1758, Alcalde Mayor Primo de Rivera reported paying out 4,090 pesos and 7.5 reals for Totonac laborers, which equates to 32,727.5 peonadas, or approximately 90 workers per day over a 365-day period.[4] For the preceding year, he paid 5,449 pesos and 0.5 real for 43,592.5 peonadas, an average of approximately 119 workers per day.[5] To better standardize comparisons, I used a full year of 365 days to calculate the preceding average, but work did not proceed on each and every day of the year. The typical rural laborer actually worked only about eight months out of a typical year.[6] As shown in a later section in this chapter, road labor seems to have continued with fewer interruptions, but in eighteenth-century Mexico the rhythm of work still had many breaks in it. One historian finds many interruptions of the work routine by miners in Chihuahua. Holidays, festivals, and religious observances combined to reduce substantially the time spent working in the mines. As the century progressed, mine operators and governors worked to reduce the number of public holidays in order to increase the miners' production.[7] One would expect similar practices concerning holidays for the gangs working on the roads around Xalapa, but on this point the records of 1757–58 remain silent.

Villagers used different approaches to the problem of supplying these manpower requirements, and the records show a case of one pueblo changing its method of distributing the burden from one year to the next. The small village of Chiltoyac sent nineteen men to work for 171 peonadas in 1757, or 9 man-days of work from each villager sent. In the following year, thirty-five villagers from Chiltoyac performed 147 peonadas, or about 4 man-days of work from each.[8] For some unrecorded reason, the load was spread among more of the villagers in 1758 than had been the case in the preceding year.

To assess the impact of the labor demands on the entire village, one can estimate the population of Chiltoyac by extrapolating backward in time from the census taken of the diocese of Puebla in 1777. The manuscript records that Chiltoyac included fifty-eight households, seven of which were headed by widows.[9] The findings of Sherburne Cook and Woodrow Borah for western-central Mexico show that the indigenous populations there increased by approximately 50 percent over those same twenty years.[10] If the Totonac around Xalapa recovered their numbers at a comparable rate, then Chiltoyac in 1757 would have consisted of some forty to forty-five households. In any event, the number of villagers eligible for draft labor must surely have been less than fifty. The nineteen men who averaged nine days of work in 1757 thus constituted a sizeable fraction of Chiltoyac's available workforce. Bringing thirty-five men in the following year reduced the burden to four or five days per capita, but also would have required the mobilization of virtually every eligible male in Chiltoyac for about one week to fulfill the colonial regime's labor demands.

The demands of the Spanish colonial regime on these Totonac villages came in a number of guises. There was also tribute to be paid. Analyzing the intersection between village tribute and the alcalde mayor–arranged repartimiento of goods, one historian finds considerable manipulation of those systems by the Mixtec and Zapotec of Oaxaca, challenging earlier scholarship that emphasizes their victimization.[11] Around Xalapa, we find a somewhat parallel situation as Totonac villages performed extra labor repartimiento service in order to satisfy tribute demands.

The pueblos of Coatepec (the one near Xalapa) and Xico provided labor drafts more than twice as large as any other place, including the barrio of Xalapa itself. In both 1757 and 1758, some of the Coatepec and Xico villagers' labor counted toward partial satisfaction of their annual tribute requirement. As can be seen in table 3.1, 626 Coatepec villagers worked 2,672 peonadas during 1757, and an unspecified number of persons worked 2,383.5 peonadas in 1758. The comparable figures for Xico are 3,260 and 3,440.5 peonadas in those years, respectively, and the reports merely state that all of the villagers from Xico performed draft labor on road projects.[12] Xico's obligation increased, whereas that of most other villages decreased, but the omission in the report of any specific number of workers for either year precludes calculating how much of the Xico villagers' road labor went against their tribute requirement.

Table 3.1. Draft Labor Supplied by Various Pueblos around Xalapa

Village	1757			1758		
	persons	man-days	payment	persons	man-days	payment
Coatepec	626	2,672	334 pesos[a]	?	2,383.5	297 pesos, 7.5 reals[a]
Chiltoyac	19	171	27 pesos, 3 reals	35	147	18 pesos, 3 reals
Jilotepec	"todos"	1,172	146 pesos, 4 reals	?	1,442	177 pesos, 6 reals
Perote	30	180	30 pesos	?	144	18 pesos
Xalapa	"todos"	1,114	139 pesos	275	953	119 pesos, 1 real
Xico	?	3,260	407 pesos, 4.5 reals	?	3,440.5	430 pesos, 0.5 real[a]

Source: AGN Caminos y Calzadas, vol. 5, ff. 136, 323–31.
[a]Payment discounted to account for tribute remission.

Coatepec's figures, however, include details that facilitate an estimate of how much of the villagers' labor was performed in satisfaction of the labor draft and how much was performed in lieu of the pueblo's required tribute submission. Because the areawide reductions in draft labor for road construction between 1757 and 1758 were 25 percent, four times the difference in peonadas worked may represent the number of man-days required to satisfy the demands of the labor draft itself. Subtracting that figure from the total would leave the number of peonadas worked in lieu of tribute. In the case of Coatepec, the result is 1,518 peonadas. At a single real per peonada, Coatepec would have earned an extra 189 pesos and 6 reals to be applied against the tribute requirement. Primo de Rivera's reports nowhere indicate that the two villages only partially fulfilled their annual tribute demands by substituting road labor, but this was indeed probably the case.

These road-construction expense records contain no statement of the monetary value of the total tribute requirement of Coatepec or Xico, so it is impossible to state definitively whether these two pueblos met their total requirement by substituting road labor. If one assumes that tribute was demanded in magnitudes similar to that for native pueblos elsewhere in

New Spain, then the problem becomes somewhat simplified. Gibson found that the per capita tribute requirement in the late eighteenth century ranged between 13 to 22.5 reals, with most individuals paying 17.5.[13] The average stint of a Coatepec laborer was 4.26 peonadas, somewhat less than the range of 5 to 6 noted earlier. The 626 individuals must therefore have represented a large fraction of Coatepec's tributary population, indeed if not all of it. In any case, using the lowest of Gibson's range of values, these 626 tributaries were forced to pay 1,017 pesos and 2 reals each year. Gibson also notes that in 1771 José de Gálvez reported a range of 8 to 24 reals per tributary, but even this lowest rate would set the requirement at 626 pesos, still much more than the tribute estimate of 189 pesos and 6 reals derived from the data, which would have met only a part of Coatepec's annual tribute requirement. The documents do not reveal exactly how the Totonac governors of Coatepec and Xico arranged with authorities to use road labor to meet part of their tribute requirement. Some of these colonial authorities, however, appear to have grown increasingly dissatisfied with the results of the village labor drafts, as was the case with Primo de Rivera.

It is unlikely that the Xalapa laborers were trying to slow down production so that they could remain on the Crown's payroll for a longer period. As shown later in detail, one real per day was a very low wage for the period, barely above the subsistence level for a single wage earner. More likely the reluctance was itself a form of passive resistance to forced labor—in other words, a strategy for coping with the colonial regime's labor demands.

How great was the impact of the road labor repartimiento on the various pueblos? This is a difficult question. One week or so out of an entire year may not seem like much at all. However, even a short length of time could have a significant impact if agrarian producers were summoned away from their fields at important points in time. A comparison with a parallel situation in Japan during the Tokugawa shogunate (1603–1867) is instructive. Constantine N. Vaporis analyzes the burdens placed on rural villages to service the extensive official highways that connected major cities and important pilgrimage destinations. These roads were marked by stations separated by approximately one day's journey on foot. Assisting villages were designated for each station and then assigned a variety of tasks to support the station. Vaporis finds that villages used a variety of means to resist the demands that the *bakufu* (shogun's government) placed on them. Peasants complained that their stints of several day's service had a significant and deleterious impact on the welfare of their villages. Vaporis concludes that the burden of a mere few days of labor each year was significant to each Japanese peasant who had to do it. In addition to resistance, the Japanese villages also negotiated their burdens with the government. Some assisting villages were able to escape the requirement to supply manpower by substituting cash payments.[14] In a sense, Coatepec and Xico did something similar by negotiating to supply extra labor in order to escape some of their tribute burden.

Alcalde Mayor Primo de Rivera's reports for the years 1757 and 1758 were submitted in the early months of the subsequent years. They give no indication as to exactly when the Totonac villages had to send their labor drafts. The letter to Viceroy Ahumada in which Primo de Rivera gave vent to his frustrations was dated October 25, 1757. And in a letter to Primo de Rivera on September 7, 1757, Miguel Cárdenas (apparently the director of all road construction and repair in the area) reported that 3.55 kilometers (2,893.5 varas) of paved road had been completed in several locations, including Cerro Gordo, Xalapa, and the pass through the mountains leading to Perote and on to Puebla. Also, work continued on the two bridges mentioned earlier.[15] One can logically infer that the alcalde's ire had been building for some time, so the work demands must have been made during the growing season. The impact on the Totonac peasants cannot have been negligible. A modern-day highway atlas records the distance between Xalapa and Veracruz as 115 kilometers, and between Xalapa and Mexico City as 306 kilometers. The artist Carl Nebel created the early-nineteenth-century lithograph seen in figure 3.1, which shows the eastern approaches to Xalapa. The Camino Real obviously had a long way to go to become a paved connection between Xalapa and Perote, let alone between Veracruz and Mexico City. The expense of such a project would have been horrific, so it seems likely that all that could be reasonably afforded were improvements at sites of importance to local populations as well as at strategic bridges and mountain passes.

TRANSITION TO FREE-WAGE LABOR

A series of weekly payroll records from 1767 through 1769 shows a significant change in how the colonial regime recruited day laborers for the Camino Real. We find that road work continued in many of the same locations a decade after the period covered by the first series of records. Road

Figure 3.1. Xalapa, Lithograph by Carlos Nebel. This print, based on the artist's visits to the area in the 1830s, shows what the general appearance of the eastern approaches to Xalapa would have been thirty years earlier. The Cofre de Perote can be seen beyond Xalapa in the background. COURTESY OF COPLEY PRESS.

crews were at work at San Miguel del Soldado and Banderillas on the road leading west from Xalapa through the mountain passes to Perote and at Las Ánimas, which lay immediately east of Xalapa and is now a part of the city itself.[16] If these surviving records are complete for their reported periods of time, then four crews worked simultaneously, apparently at sites usually quite close to one another.

The earlier scheme that levied draft requirements on whole native villages had been abandoned in favor of a free-wage labor regime. The documents submitted from 1767 to 1769 are organized in a much different manner than those prepared during 1757 to 1758. Instead of summaries by pueblo and barrio, line entries appear that name each individual who worked in a given week, along with a notation of how many days that person worked. Village draft quotas are replaced in these records by rosters of persons usually listed as *jornaleros*, but sometimes alternately as *trabajadores* or even as *labradores*, which generally translates to "farmers." These day laborers received two reals per day, whereas the previous scheme had paid only one real for each draft worker demanded by the alcalde mayor. As discussed in a later section in this chapter, two reals per day was above the subsistence level for one person, but would probably not support more than one additional person for any length of time.

The results from the 1767–68 Xalapa records indicate not only that the weekly turnover rate of the day-laborer force remained high after wage labor was introduced, but also that although day laborers were very unlikely to reappear in a subsequent week's report, only rarely did any of these individuals fail to finish out a full workweek for which they were hired. Each name appears alongside an entry that notes the number of days for which that individual was paid. As one would expect because of holidays and weather delays, the crews worked more days in some weeks than in others. In only 2.9 percent of the entries is a laborer reported as having worked less than the maximum number of workdays for the particular week. As seen in chapter 4 in the analysis of a project undertaken much later in the eighteenth century, this "absentee" rate is quite low. Obviously, one must not summarily dismiss the possibility that the overseer simply failed to report an absence and then pocketed the difference in pay. Except for Las Ánimas, most of the construction sites in 1767 and 1768 lay sufficiently far from Xalapa that it would have been difficult for the alcalde mayor to approach unnoticed and catch a discrepancy between the number of laborers present in the flesh and on paper. However, those few cases of workers whose wages totaled less than the full week's amount show that at least some of them left the crew when they probably could have stayed on to earn a few more reals.

Despite the change from draft village labor to a free-wage labor regime, it appears that indigenous laborers were still the ones who actually performed unskilled work on the roads and bridges. As seen in chapter 4, by 1791 creole and mestizo faces would frequently appear among the unskilled laborers on construction work, but in 1767 this workforce was still overwhelmingly composed of the Totonac. Because there is no census manuscript exactly contemporaneous with the payroll documents, it is very difficult to determine biographical information about the individuals listed.

Only one of the many payroll reports identifies the *calidad* (official "racial" classification) of the individuals who appear on it, but on that document the sobrestante did in fact list the day laborers under the category "yndios trabajadores."[17]

The same people did the work, but direct hiring of individuals for the crews removed the Totonac village governors from their crucial role in the colonial draft-labor regime. In an individual sense, the Totonac day laborers may have benefited from personally receiving the daily two reals. In a collective sense, however, undercutting the village governors' authority must have in some measure weakened the fabric of the república de yndios. Since the Conquest, Spanish colonialism had relied on native elites physically to marshal the indigenous labor force to work at a variety of tasks. Robert Haskett finds that the native elite in Cuernavaca maintained its social importance to the end of the seventeenth century through a variety of means of coping with colonialism. Native governors in that area worked for years to rid their towns of the repartimiento requirement for labor in the Taxco and Cuautla mines. They finally succeeded in 1755. Haskett contends that the office of governor retained its importance throughout the eighteenth century and past independence.[18] Steve Stern contends that differences between various Mexican indigenous peoples resulted in regionally different colonial relationships with Spanish officialdom at various levels.[19] Antonio Escobar Ohmstede analyzes the changing roles of *cabildos* (city or town councils) in indigenous pueblos around Veracruz.[20] A wealth of literature addresses the crucial matter of indigenous officials throughout colonial Spanish America in times of relative calm and during massive rebellions. Space requirements limit me here to name but a few of the many important studies on this subject: James Lockhart and others on the Nahua of central Mexico; Kevin Terraciano on the Ñudhzahui (Mixtec) of Oaxaca; Nancy Farriss, Robert Patch, and Matthew Restall on the Maya in Yucatán; and Karen Spalding and Steve Stern on the indigenous peoples of Peru.[21] Charles Gibson analyzes in detail various mechanisms utilized throughout the colonial period to collect tribute and emphasizes the important role played by the native governors as the conduits for the range of demands on their villages, including not only tribute collection itself, but also repartimiento and other manpower levies.[22]

Lockhart finds that as the centuries passed, the Nahua experienced greater acculturation to Hispanic influence and no longer depended so strongly as before on indigenous officials in dealing with the Spaniards. Similarly, Terraciano concludes that by the eighteenth century Spanish colonialism no longer depended on the *yya* (caciques) in dealing with the Ñudhzahui. One sees clearly that in Xalapa by the late 1750s this system had become a two-way street. Alcalde Mayor Primo de Rivera felt that he could more effectively mobilize indigenous labor if he could bypass their governors. Ten years later Primo de Rivera's successors were doing exactly that. It would, of course, be ludicrous to see in this one change the complete demolition of indigenous governors' power. One must concede, however, that when those governors were stripped of this particular function, they suffered an incremental diminution of power and possibly status. The erosion of these individuals' collective status would not have occurred all at once. The change more likely resulted from an accumulation over time of small changes such as this one.

Women undoubtedly composed a large fraction of the labor force in New Spain, as shown in a study of their role in royal tobacco factories such as the one in Orizaba.[23] The reports from the various projects around Xalapa in 1756–59 and 1767–70, from the Puente de Escamela project in Orizaba in 1791–92, and from the 1794 Toluca road construction analyzed in subsequent chapters contain no record that any woman ever worked on them. The absence of any women's names from these records offers a rather startling contrast with sixteenth-century French practices that perhaps also existed in Spain. Frenchwomen were often employed as day laborers on municipal construction projects and earned only one-half to two-thirds as much as did their male counterparts.[24] Although the archival records contain no specific documentation, it seems likely that women were present in significant numbers among the labor drafts from Totonac villages. When most of the men left a pueblo such as Chiltoyac to perform road labor, some women villagers might have accompanied them, somewhat like camp followers, and acted in a variety of roles even if their names did not appear on the official payrolls. Primo de Rivera specifically complained that indigenous governors had been charging the Crown for women villagers. The distances were not great, however, so the absences were likely not long. It is also possible that women villagers remained in their pueblos to carry out their part of the gendered division of labor.[25] Future research may unearth evidence that will give voice to the currently silent record.

The transition to the use of unskilled, free-wage labor obviously did not occur all at once throughout the viceroyalty. Examples of the use of drafted laborers surface in years after the 1767–69 projects around Xalapa. The Totonac governor of Jilotepec complained in 1780 about being required to grow tobacco while still having to meet requirements to supply road-construction labor and materials. These latter demands had been levied since 1771, shortly after the work records analyzed earlier.[26] It is easy to see why cabildos might prefer to draft labor at one real per day instead of hiring free labor at twice the draftee wage.

Some very interesting correspondence during 1776 suggests that colonial officials at the highest levels may to some degree have forced the transition to free-wage labor on reluctant local governments. The Ayuntamiento de Puebla requested that Viceroy Antonio María de Bucareli authorize a repartimiento labor draft to repair a road bridge in that city. Because the bridge linked Puebla with Mexico City, it obviously was a very important structure. The Puebla officials proposed not to pay the workers any cash at all. Instead, the laborers were to receive tools and meals in the time-honored tradition. Viceroy Bucareli rejected the request and directed the use of hired wage laborers. In his curtly worded reply, he ordered the Ayuntamiento de Puebla to get on with the repairs to the bridge and to complete them quickly and economically in keeping with the principles of good government.[27] It is worthy here to note that local officials in the Cuernavaca area made a similar attempt to reinstitute the Taxco mine labor repartimiento.[28]

The Ayuntamiento de Puebla's request for a repartimiento is easy enough to understand. Forced labor at no pay is profitable for any employer, as perhaps most eloquently explained by Alfred H. Conrad and John R. Meyer, whose econometric analysis convincingly demonstrates the great profitability of U.S. slavery in a system of capitalist, commercial agriculture.[29] Although the ayuntamiento was not a profit-making enterprise, it certainly must have wished to minimize its expenditures in completing public works, so it submitted a petition for a forced labor draft.

Wages and the Labor Supply

When required to rely on free-wage labor, governing officials had next to solve the problem of how to attract it. In eighteenth-century New Spain, a common solution was for an employer to pay an advance of some combination of cash and goods. Once indebted, a worker would then be bound to his employer until the debt was paid. Many historians of colonial Mexico have studied this arrangement, often referred to as debt peonage. Cheryl English Martin recently analyzed its importance throughout the eighteenth century as a solution to the problem of low wages and high prices that faced miners around modern-day Chihuahua.[30] Eric Van Young emphasizes the role that debt peonage played in stabilizing the labor force throughout New Spain during this period and finds that it was a pervasive and even central aspect of the labor market throughout the realm, even though its per capita importance declined in the Guadalajara region toward the end of the eighteenth century. In Guadalajara, a hacienda peón earned approximately four and one-half pesos per month in the mid-1780s.[31] William B. Taylor finds that a debt peón in eighteenth-century Oaxaca earned a monthly wage of approximately three and one-fourth pesos and concludes that debt peonage may have afforded a "strong bargaining position" to the agricultural laborers of the Valley of Oaxaca.[32]

The free-wage day laborers on the Xalapa projects received two reals per day, consistent with the common daily wage earned by "most hacienda peons," according to Charles Gibson.[33] The Oaxaca agricultural laborers received between twenty-six and twenty-seven reals per month, which is approximately one real per day if some allowance is made for Sundays and holidays. This Oaxaca figure amounts to a bit less than the rate of thirty-six reals per month that prevailed in Guadalajara, but still falls just short of two reals per day paid in Xalapa. Although one can reasonably surmise that hacienda peons also received rations-in-kind, the records indicate only that the road workers were paid in cash. Payment in coin at a rate higher than the daily rural wage most likely eliminated any need to attract workers with an *enganche* (down payment or advance).

Van Young finds that debt peonage was normally associated only with permanent employees, not with temporary workers such as the Xalapa day laborers in 1767–68.[34] The skilled labor force also had no debt component evident in its records, and although the format of the records indicates that they were hired on a weekly basis, enough of them returned week after week that they constituted a fairly stable workforce. They received daily wages considerably higher than did the day laborers, so there was no need to entice them with advance income in the form of debt.

Here perhaps lies at least part of a possible explanation of the apparently irregular work habits of the many day laborers who worked one or two weeks on the road and then disappeared from the project payrolls. These men were most likely agricultural workers who left for short periods to work on the road crews. Workers on agricultural estates were paid by the month, but they likely did not work every day of a month.[35] On a road crew, they could supplement their important payments-in-kind with a little hard cash. Because tropical vegetation would have grown quickly between recently laid paving stones and would render unusable the roadbed and its adjacent clearings, road work was almost certainly regularly available to these workers. One can assume that most members of their families worked at something and that many probably supplemented their agricultural payments-in-kind with vegetables from a small garden.

Road repairs seem to have continued apace irrespective of holidays and festivals. The draft-labor records from 1758–59 contain no mention, one way or the other, of work stoppage for holidays and religious festivals. It would be a surprise to find that the draft laborers were allowed to ignore these important ceremonies. These rituals almost certainly served an important role in social control, as Martin found they did in Chihuahua. There, lengthy shutdowns occurred, especially around the Christmas holidays.[36] The labor records from 1767–69, however, indicate that few if any delays were accepted for these observances. During three weeks in April and May 1767, the day-laborer crews worked a full seven days, and in a fourth consecutive week they received pay for six and one-half days.[37] During the week including Christmas of 1767, each of the four road crews received six days' wages.[38]

Artisan Road Laborers in Xalapa

The successful completion of each of the construction and repair projects examined in this study must, of course, have owed a great deal to skilled artisans, even though unskilled day laborers composed the major portion of the workforce in each case. *Albañiles* (stonemasons) actually placed the stones to build the Puente de Escamela and laid the pavement around Xalapa and on the Mexico City–Toluca road. One historian of roads states that in the eighteenth century stone "paviers" graded stones to a standard size and then laid them to form the paved surface. Supported by an "army of stone breakers, [t]he pavier . . . was something of a skilled artisan."[39] Such tradesmen were always hired individually on a free-wage basis, so we find that in 1757, when village drafts provided the army of unskilled workers for the repair projects around Xalapa, the alcalde mayor also hired an array of skilled laborers, including two stonemasons, seventeen stone-pavers, and five carpenters. Chattel slavery, repartimiento labor, and free-wage labor existed simultaneously in Tehuantepec since the sixteenth century, so it comes as no surprise to find two different labor regimes still employed around Xalapa in the middle of the eighteenth century.[40] The 1757 documents show forty-four wage laborers employed at the various Xalapa sites: a director, two stonemasons, nine sobrestantes, four *oficiales* (whose function remains obscure), five carpenters, seventeen stonepavers, including four masters and thirteen *media empedradores* ("half" or jour-neymen stonepavers), and six *peónes a mano* (laborers whose trade and function remain unspecified).[41] As shown in table 3.2, their wages ranged up to one peso per day. The absence of week-by-week records precludes the detailed analyses of employment patterns made possible by the later records, but it is reasonable to assume that the 1757 trends paralleled those of artisans who appear in records from 1767 and 1768 and who also worked under a free-wage labor regime.

Table 3.2. Daily Wages on Road-Construction Projects in Late Bourbon Mexico

	1757 Xalapa	1767–68 Xalapa	1791–92 Orizaba	1794 Toluca Road
Overseer	1 peso	1 peso	6 reales	
Stonemason	1 peso[a]	4 reales	5 reales	4 reales
Carpenter	4 reales[b]	3 reales		
Paver (empedrador)		6 reales		
Paver (media empedrador)		4 reales		
Day laborer (peón a mano)	3 reales	3 reales		
Day laborer	1 real[c]	2 reales	2.5 reales	2 reales 1.5 real

Sources: AGN Caminos y Calzadas, vol. 5, ff. 380–82; vol. 8, ff. 151–364, and vol. 14, ff. 184–360; AGN Peajes, vol. 1, ff. 52–195.

[a]One person received nine reales; the other received seven reales.
[b]One carpenter received one peso per day.
[c]Several thousand drafted yndio village laborers.

By the time free-wage day laborers replaced the repartimiento workforce in 1767, the stonemasons and stonepavers had come to be classified simply as albañiles, a practice followed in the records from the Puente de Escamela and the Toluca road. As shown later, skilled laborers also suffered economic blows as the labor market of New Spain tightened during the last decade of the eighteenth century.

The persons in higher-paid labor categories exhibited work patterns that differed significantly from those of the day laborers. Sobrestantes were of course present, supervising their own crews each week, and in 1767 they still earned one peso per day. The largest number of artisans were the albañiles, or stonemasons. Twenty-eight of them worked on the various Xalapa road projects between November 16, 1767, and August 24, 1768, along with one *mesclero* (mortar mixer) named José Antonio, who moved from crew to crew. This is a larger number of stonemasons than in the 1757 projects, but the 1767–68 reports mention no empedradores. The individuals who would earlier have been listed as stonepavers were apparently now lumped in with the albañiles.

The stonemasons earned four reals per day on the job, and the longest-working stonemason earned seventy-one pesos during the thirty-five weeks under consideration. The mortar mixer earned two and one-half reals per day, a wage that rose to three reals by 1769. These records show a strong propensity of the stonemasons to work for long stints of several weeks each. Five of them appear in more than half of the thirty-five weekly records, and many of the others whose employment was shorter-lived often worked four or five consecutive weeks before moving on.

The stonemasons appear to have been mostly Totonac residents of the barrio of Xalapa. A comparison of their twenty-eight names against the 1777 census list yields sixteen name congruencies. This seems an exceptionally high number, especially because the documents are separated in time by more than a decade, in which case death or migration may have eliminated some of the possibilities for name-matching record linkages. In stark contrast, only four name congruencies resulted from a comparison of the nonyndio census and the list of stonemasons' names, and two of those were names that also appeared on the yndio list. Stonemasonry appears to have been the province of indigenous workers in 1767, just like unskilled day labor.

Historical Comparisons and Conclusions

One way of interpreting the significance of the labor regime in the eighteenth century is suggested by Enrique Semo in his study of the development of capitalism in Mexico. Semo finds that the república de yndios acted as the principal structure in a "tributary despotism" mode of production, fundamentally distinct but inextricably intertwined in colonial society with a feudal *república de españoles* that also exhibited elements of "embryonic capitalism." Importantly, the tributary despotic mode of production in the labor system contained extraeconomic elements of coercion.[42] Whether paying tribute or serving in a road labor draft, the indigenous peasant "contributed" a surplus that was extracted through political means. Here, the Mexican situation seems to parallel the more general "tributary mode of production" suggested by Eric R. Wolf.[43]

A more universal framework is suggested by Samir Amin, who contends that the worldwide capitalist mode of production was preceded by a "family" of tributary modes. In Amin's scheme, European feudalism is one such mode and is unique because of its political decentralization.[44] Semo, Wolf, and Amin all offer variations on a theme—a deterministic view of history where the human world progresses through modes of production and arrives at the capitalist phase.

We might see Viceroy Bucareli's insistence that the Ayuntamiento de Puebla use free-wage labor for bridge repairs in a similar, deterministic vein. Free-wage labor is a sine qua non of the advent of capitalism. The viceroy's reply in a sense speaks for itself: he told the Ayuntamiento de Puebla to hire free-wage laborers because that was what a good government should do. One historian finds that Bucareli was essentially a conservative administrator, but notes that at the same time he served as an agent of Enlightenment ideas during his tenure as viceroy of New Spain during the so-called Bourbon reforms.[45] Of course, in a Marxian view, the advent of the Enlightenment was an essential part of the capitalist revolution that ushered out feudalism.

In order to assess better any theoretical implications of the transition from drafts to free-wage labor in eighteenth-century New Spain, we might step back and view a broader sweep of time and space. The Spanish colonial regime's use of repartimiento labor was by no means something invented to accompany its conquest of the Americas. Instead, repartimiento labor was a centuries-old custom.

The Roman Empire is legendary for its road network. Over a period of centuries, a web emanated outward from Rome through Italy and spread around the provinces. Major Roman roads were generally built for military purposes. The troops of the legions could move quickly on them to reach a border province or the site of a local rebellion. Following the military reforms of Gaius Marius in 106 B.C.E., construction of major public roads became the work of the legions' soldiers. A shovel became part of each soldier's standard equipment. Road building had thus become a task performed by paid employees of the state, although delays might result if combat requirements took priority for the soldiers' services. Professional engineers and architects usually supervised construction, and authority existed to conscript local laborers should a situation warrant. Once the roads were built, the responsibility of maintaining them was left to local communities. Prisoners were frequently used for road maintenance, and towns and villages often used corvée labor to fulfill their upkeep obligations. Historians differ over how well local governments maintained Roman roads, but considering that such an extensive network existed over several centuries of time, those who contend that many roads often suffered from neglect seem to have reached a more logical conclusion. The principal source of road-maintenance labor appears to have been corvées from villages along the routes, but in the province of Hispania there survives a record of the existence of a guild of road-surface workers in the present-day city of Tarragona. A guild implies the existence a group of persons who work full-time at something, instead of the rotating, part-time work that characterizes a corvée system. So from time to time there were exceptions to this general Roman practice.[46]

Skip ahead several centuries to Rome's successor, the Byzantine Empire, and on to that empire's Ottoman successor. The Turkish janissaries did not build roads like the Roman legionaries; instead, construction burdens were thrust on villages that continued their earlier maintenance duties under a corvée system. The Ottomans frequently conscripted nomads for construction, but settled villages bore the burden of upkeep. Sometimes this burden earned them an exemption from some other tax burden, a situation somewhat akin to the one negotiated by Coatepec and Xico in New

Spain.[47] Another common Ottoman practice echoed the situation of the assisting villages along Japanese highways.

Key points along Ottoman roads were garrisoned by soldiers known as *derbent* guards. Derbent guard posts were maintained by nearby villages, which also bore all associated expenses. Local streets, roads, and bridges, in contrast, were maintained by local governments, usually under the cognizance of a magistrate called a *qadi*. This official was a member of the *ulema*, a local council of Islamic religious leaders. Although under the Ottoman system each religious minority had its own local government or *millet*, the situation evokes very much that between an alcalde mayor and the cabildo, or town council, in Mexico.[48]

Prior to the Spanish conquest of the Americas, exaction of corvée labor by the state was in places a significant source of revenue and service. In Mesoamerica during the Late Formative Period (ca. 500 B.C.E.–200 C.E.) corvée labor was used to construct the temples of Oaxaca, and some contend that in general the mound-building burden on the populace was not very great. But more than a millennium later the Mexica-Aztec Empire made little use of corvée labor. The *tlatoque* ("emperors") of Tenochtitlán found it convenient to allow a great degree of local autonomy to subject *altepetls* (city-states), which tended to militate against imposition of corvée requirements.[49]

In contrast, the Inka rulers made extensive use of the corvée throughout most of Tawantinsuyu, their empire in today's highland Peru. In a classic work that established the direction of research on the subject, John V. Murra found that the Inka state relied heavily on corvée labor in both the production and distribution of subsistence goods as well as in public services. Other scholars have found considerable variation in the assignment of Inka state burdens. Of particular relevance to this study is Terry Yarov LeVine's findings that some Andean communities were assigned road-maintenance duties and that the burden seems to have been somewhat unevenly applied.[50]

One can see thus far that although the repartimiento labor system had deep historical roots, it was by no means peculiarly Spanish and not even limited to Eastern Hemisphere societies. Similar practices had emerged in the Americas before the European invasion. The Americas had been isolated from the Eastern Hemisphere for several thousand years before undergoing the agricultural transition from hunters and gatherers to herders and cultivators, a crucial step in the evolution of complex societies such as Tawantinsuyu, imperial Rome, or China.

China is legendary for its Great Wall and Grand Canal, as are Tawantinsuyu and imperial Rome for their road systems. Its history adds another dimension to a study of public-works labor. Historians of China periodize their study by dynasty. During the Qin and Han dynasties (221–207 B.C.E. and 207 B.C.E.–220 C.E., respectively), the first of the great walls was built by a combination of troop labor and corvée labor.[51] In the early Han dynasty, a system of compulsory labor service for public-works maintenance was instituted. Males between fifteen and fifty-six years of age owed one month of unpaid service per year on public projects and were also liable for an entire year's service once in their lifetimes. In the later Han period, there had been "a partial transformation of the one-month labor service obligation into a monetary tax." This was probably encouraged by district magistrates, who found it simpler to hire peasants and pay them than to apportion and administer a draft-labor service.[52] Recall that Alcalde Mayor Primo de Rivera did not make a parallel argument to Viceroy Ahumada when arguing for substitution of wage labor for repartimiento labor for the roads around eighteenth-century Xalapa.

Most of the Grand Canal was originally built during the short-lived Sui Dynasty (581–617 C.E.). The Yang-ti emperor compelled his subjects to meet heavy demands for corvée service in constructing this wonder of the world. One million men were conscripted for one section, and one hundred thousand for another. Chinese peasants offered considerable resistance to the demands on their time.[53] Both the successor Tang and Song dynasties retained corvée labor requirements, and each at different times instituted a method for those who could afford it to buy out of their obligation by paying a tax that in theory would be used to hire a substitute.[54] In the Ming Dynasty (1368–1644), all roads were maintained by locally administered corvée labor, as was the Grand Canal. There was no subsidy from the central imperial government.[55]

Some major attempts at reform occurred during the last dynasty of imperial China. During Qing times (1644–1911), the Kangxi emperor decreed in 1713 an end to the household-based corvée labor system that provided labor for public-works maintenance. In an attempt to increase productivity, hired laborers would perform those tasks, and the imperial regime replaced the corvée demand with a head tax based on the number of adult males in each household.[56] It is unclear what happened as Qing rule unraveled in the nineteenth century, but it seems less likely that a funded system of labor recruitment would have been administratively feasible, especially in view of the magnitude of upheaval brought by the Taiping rebellion.

Corvée labor reemerged in Yucatán during the eighteenth century. A system of labor drafts called *mandamientos* was established to mobilize the labor supply to fill a variety of demands, specially hacienda labor. Mandamientos were under the control and jurisdiction of the subdelegate, or the magistrate serving as the local executive power.[57] Thus, the Spanish colonial regime reinstituted a corvée in one part of New Spain at the same time it was insisting on the use of free-wage labor elsewhere. Mandamiento labor was used to build a stone bridge near Mérida in the late eighteenth century. Beginning in 1787, Maya labor was organized into gangs of twenty at a time to build the Potoktok Bridge.[58] Mandamientos persisted in Guatemala well into twentieth century.[59] Like all corvée systems, the repartimiento of the Spanish colonial regime resists any attempt to associate it clearly with only one mode of production.

Corvée systems continued to exist after the Industrial Revolution heralded the flowering of industrial capitalism. In Peru, they were the cause of some Andean peasant rebellions in the late nineteenth century. Much like the Totonac governors of the pueblos around mid-eighteenth-century Xalapa, Tuutsi subchiefs of late-nineteenth-century Rwanda implemented a variety of exactions for the German colonial regime, among them corvées for road labor. In Vietnam, resistance to corvée requirements was one of the attractions of the communist-led revolution against French colonial rulers that began in earnest in the 1930s.[60] In the Dominican Republic, a national corvée requirement was instituted by the dictator President Rafael Leonidas Trujillo Molina after he took power in 1930. Richard Lee Turits sees this requirement as an example of political and national cohesion, to a degree never achieved and apparently unachievable in the country's preceding years. One should bear in mind that the eighteenth-century Intendancy of Veracruz, not established until 1786 and coterminous with today's Mexican state of that name, has an area half again as large as the entire Dominican Republic. In the nineteenth century, local Dominican administrations had similarly allowed public labor to be substituted for payment of road taxes. Under the Trujillo regime, like the subjects of the Song and Tang dynasties in China centuries earlier, Dominicans had the option of paying a tax in lieu of working a few days each year.[61]

A comprehensive historical analysis of corvée labor across time and space has yet to be written. Such a project, however worthwhile, is beyond the scope of this study. It is useful to conclude by suggesting some tentative hypotheses to be tested by future research. It is perhaps also useful to note briefly the extensive historiographical debate over the nature of the Mexican hacienda during the colonial period. Although Spanish interest lay principally in silver mines, most Mexicans worked in agriculture. Annaliste historian François Chevalier set the baseline in 1952, when he asserted that the hacienda had a feudal nature. Years of study of the hacienda in different Mexican locations followed, and a host of subsequent scholars concluded instead that the hacienda was essentially capitalist.[62] Attempting to sharpen the focus and to clarify the matter, Steve J. Stern discusses the many problems with applying either the feudal or the capitalist model to colonial Latin America, suggests the importance of social relations of production rather than markets in analyzing modes of production, and concludes that an entirely new model is needed, entitled perhaps a "colonial mode of production."[63]

We have seen that corvée labor systems such as the repartimiento emerged in societies separated by oceans of water and by millennia of human history. With the exception of the supposedly socialist states of the twentieth and twenty-first centuries, we have seen that these corvée systems existed in all of the modes of production defined by Marxist historians and by those of the Annales school. We have seen them sometimes reappear in situations where they had once been replaced with free-wage labor. Because corvées are government exactions of temporary labor, by their very nature they are always backed by Semo's extraeconomic elements of coercion. The corvée is essentially a form of taxation. As long as there is a government that has sufficient political hegemony to enforce the corvée and that is not somehow prohibited from using such power, the possibility of some sort of corvée exaction is always present for that government's public works. Some regimes may have found it preferable to dispense with the administrative complications involved in assigning the burdens and mobilizing corvée labor forces and conceivably also to avoid possible resistance by villagers who felt unduly burdened. It does seem logical that expansion of the world cash economy that attended industrial capitalism would thus make use of a corvée less likely than in earlier centuries, and one should keep in mind that the Dominican Republic's government in the early twentieth century was perennially short of cash. If a regime has the cash and a supply of workers who are sufficiently integrated into the economy to be willing to accept it, offering a free wage to attract the necessary workers provides a relatively simple solution for a government to achieve its ends.

Demographic matters are treated in chapter 5, but suffice it here to note that the population of late-eighteenth-century Mexico had steadily increased up until around 1770.[64] If economic coercion failed to work, it could be replaced by administrative coercion. In other words, should the wage prove unable to mobilize the necessary labor supply, the compulsion option always remained, although the next chapter shows that the Spanish colonial regime did not use the repartimiento option for projects near Mexico City and around Orizaba during the last decade of the eighteenth century or in the years immediately following the turn of the nineteenth century. There was no need for the viceroy to do so.

CHAPTER FOUR

TIMES GET TOUGHER

THE PUENTE DE ESCAMELA AND THE TOLUCA ROAD

As discussed in the preceding chapter, efforts to improve the transportation infrastructure in Mexico continued through the last years of the colonial period. This chapter explores changes in the Mexican labor market during the final decade of the eighteenth century. The documentary evidence for this study consists of a complete, eighteen-month-long series of records of expenses incurred during the construction of a highway bridge near Orizaba and a thirteen-week-long consecutive series of similar records covering the construction of a paved highway between Mexico City and Toluca.

As was true of the 1767–68 Xalapa documents, the records from Orizaba and the Toluca road show that unskilled workers hired onto the jobs under a free-wage labor regime. The day-laborer wage rose by 1791, but inflation caused increases in the cost of living, which offset the gains. By 1794, road-construction laborers suffered wage cuts while inflation continued, which substantially reduced their real wages. The concomitant deterioration of their economic position marked another step toward proletarianization of these unskilled workers.

ORIZABA AND THE CAMINO REAL

The villa of Orizaba lies near the volcano Citlaltepetl (Pico de Orizaba) on the eastern slope of the mountain range facing the Gulf of Mexico. Orizaba's elevation is approximately twelve hundred meters above sea level, and, as at Xalapa, its temperate and rainy climate facilitates the growth of a wide range of crops.[1] During the eighteenth century, Orizaba became a center for commercial agriculture as a major supplier of tobacco for the royal monopoly.[2] Recall that the Camino Real splits in two branches immediately to the east of Puebla, with the branches reuniting in Veracruz itself. Orizaba and Córdoba are located on the southern branch, and Xalapa and Perote lie on the northern branch.

Although the physical condition of this southern section of the Camino Real appears to have been very poor throughout the colonial period, the Camino Real was very important to the residents of Orizaba, whose population rapidly increased as the city gained commercial prominence from tobacco production. One of the local priests wrote a letter to the bishop of Puebla urging reorganization of the Orizaba parishes and establishment of additional benefices so that there would be a sufficient number of clerics to tend to the burgeoning flock of both the *gente de razón* ("people of reason," meaning those persons with at least some European ancestry, no matter how unreasonable the individuals may actually have been) and the yndio population. To justify his recommendation, Doctor don Francisco Antonio de Yllueca emphasized the importance of the Camino Real and gave special mention to the new Puente de Escamela, a bridge that crossed the river of the same name where the Camino Real headed east toward Córdoba.[3] Given Fray Ajofrín's comments about the difficulty of seeing the roadbed (see chapter 2), one might question Yllueca's somewhat glowing assessment of the road because the Orizaba priest may have been motivated to embellish his description of Orizaba by visions of the heightened ecclesiastical grandeur accruing to him if his proposed expansion of the local clergy were to be approved. Yllueca included a detailed color sketch map that shows the three arches of the "new" Puente de Escamela, although its actual date of construction remains unclear. Most likely, it was one of the two bridges over which Ajofrín had crossed four years earlier.

For reasons not revealed in the records, the Puente de Escamela appears to have required replacement by 1790. Recall from the preceding chapter that Captain don Miguel del Corral, a military engineer, surveyed many public works in the viceroyalty during 1784. In his report, Corral expressed serious concern about the condition of many edifices, especially bridges that appeared to have been hastily built without competent supervision. He recommended that an ongoing effort be made to replace the hazardous structures.[4] Perhaps the Puente de Escamela fell into this category, but, for whatever reason, construction of a replacement bridge across the Río Escamela near the Barrio de Ixtaczoquitlán became a matter of priority in early 1791.

Viceroy Revillagigedo approved the project, and construction began in March 1791.[5] The correspondence that preceded the viceregal authorization has yet to surface, so the fate of the earlier bridge remains unknown. In his general history of Orizaba written in the mid–nineteenth century, Joaquín Arróniz mentions no natural calamity such as an earthquake, but he does list three other bridges that were constructed in the vicinity of the villa in the late eighteenth century.[6]

Keeping open a passable road to the important tobacco ranchos located beyond Escamela to the east and facilitating communications over the twenty-odd kilometers to Córdoba appear to have been of paramount importance to the planter and commercial elite of Orizaba. Construction of the new bridge was completed in July 1792 and was paid for using funds from toll receipts in the section of the Camino Real that fell under the jurisdiction of the Ayuntamiento de Orizaba. Although it was certainly expensive, the Puente de Escamela was very important to local markets in the two villas. Contemporaneous accounts unanimously declare that a viable commercial land artery to Veracruz itself did not exist at this point in time, a fact that underscores the importance of local market considerations to decisions regarding public works such as highways and bridges.

The Ayuntamiento de Orizaba appointed don Felipe Domínguez to oversee construction of the bridge. Domínguez received a wage of six reals for each day that he worked on the project, considerably less than the daily peso that overseers had earned in 1767. Of course, Domínguez's six reals were the highest wages of anyone directly employed in construction of the bridge.

Work Patterns of Unskilled Day Laborers

Domínguez submitted weekly reports to the Ayuntamiento de Orizaba in the same format as the ones from Xalapa during the 1760s. These reports detailed not only expenditures for materials, but also identified each individual worker, how many days he worked that week, and how much he was paid. As with reports for projects in 1757 and 1767–68, they also identified the individual workers' various job classifications. In 1791, we find that only two of the several classifications of earlier years survive: albañil (stonemason) and peón (day laborer). The first of the seventy-one reports is dated March 30, 1791, and the last was submitted on July 19, 1792.[7]

Including the overseer, at least 325 different men were employed at one time or another in building the Puente de Escamela. Those listed as stonemasons received five reals per day; 26 men earned this rate, including 4 who, in some weeks, were hired as laborers earning two and one-half reals per day. At least an additional 298 men worked as day laborers at various times.

During the construction of the Puente de Escamela, the census ordered by Viceroy Revillagigedo was prepared for Orizaba. The manuscript was approved for submission to Mexico City on November 16, 1791, so the actual enumeration would have been conducted over the preceding several months.[8] Seventy-three of the laborers can be confidently identified in the 1791 census. Many others are probably present as well, but cannot be clearly identified because of ambiguities resulting from the inability to discern which of the several possible candidates is in fact the person that Domínguez hired. Ambiguities notwithstanding, the contemporaneous Puente de Escamela pay lists and the Revillagigedo census offer opportunities for nominal record linkage rare in colonial Latin American history, and these linkages shed light on the lives of some of these humble laborers.

Record linkage using the 1791 census is complicated by the gaps in its coverage. Only a handful of those persons classified as yndios are listed in the 1791 census because they were not subject to military service. Recall that for census purposes, this study uses the juridical name from the eighteenth century, but also note that these persons were most likely Nahua, as distinct from the Totonac who worked on the projects around Xalapa.[9] These few were enumerated because they had married women of other calidades that were named and enumerated: española, *castiza*, mestiza, and *parda*. (A castiza was a woman of three-fourths European and one-fourth Indian ancestry; a parda was a woman with some African ancestry.) While traveling through New Spain on his inspection tour for the king, Fray Ajofrín noted that a 1762 *padrón* (local parish census) stated that the population of Orizaba included approximately 40 percent yndios, a fraction consistent with Susan Deans-Smith's findings for the year 1802.[10] Almost one-half of the true population of Orizaba is therefore absent from the Revillagigedo census. Because the record linkages are drawn from only half the actual population, the significance of linking more than seventy names between the two records is increased by a factor of almost two if compared to a census record that might enumerate the entire population of a locale.

As was the case with the Xalapa workers in 1767–69, a majority of the Orizaba day laborers were short-term employees who worked on the bridge for less than two weeks in any one stint. Of the 298 laborers, 179 fell into this category, of which 126 worked one or two weeks and never appeared again on one of Domínguez's payrolls. Another 65 day laborers are recorded as having worked two or more stints, including one stint of at least three weeks' duration. Seven laborers worked for between twenty-five and thirty-nine weeks. Gregorio Corona, a sixty-year-old mestizo, worked during forty-two different weeks, and Juan Estevan worked thirty-eight weeks.[11] As the latter name does not appear in the 1791 census, he likely was classified yndio and as such was hardly unique among the builders of the Puente de Escamela.

The vast majority of day laborers at Escamela appeared only once or twice on Domínguez's payrolls. Many of those who can be clearly identified in the 1791 census are listed as residing on tobacco ranchos, such as Ventura Antonio. This forty-one-year-old castizo appears in the census as an *operario* living at the Rancho del Tabaco del Sumidero and working at the bridge between June 19 and July 9, 1791.[12] Tobacco requires a great degree of attention throughout its growing cycle, with few slack periods in its cultivation.[13] Such a steady rhythm perhaps might explain why many tobacco rancho workers made only brief appearances at the Puente de Escamela, but other day laborers from the grain-producing area around Toluca showed similar work patterns. If individuals such as Ventura Antonio had only a few opportunities to leave their tobacco ranchos and pick up some cash working at the bridge, maize cultivators appeared to have had the same problem, as is shown later in this chapter.

Only 9 of the 298 day laborers worked on the bridge for more than twenty-five weeks. The actual number might be even less than 9. As with the other weekly payroll documents in this study, worker records were combined wherever possible in order to determine how many of them *could* have been exceptions to the apparent norm of only one or two short appearances by any day laborer on the Puente de Escamela payrolls. Thus, the man identified as José Antonio is recorded as having begun work stints on April 6, July 31, and November 27, 1791, followed by another one-week appearance on March 18, 1792. The 1791 government census lists no man named José Antonio, but the 1777 ecclesiastical census lists twenty-seven yndios and two mestizos by that name. Our putative José Antonio may actually have been four, or even more, different individuals. I therefore discarded the José Antonio records from the current analysis of work patterns, but this was the only name ambiguity among the Puente de Escamela documents that defied reasonably confident identification and correlation.

A few day laborers had very steady work patterns. The longest-serving peón was the aforementioned Gregorio Corona, who worked on the bridge for forty-eight of the fifty-two weeks from May 1791 to May 1792. He earned seventy pesos and seven and three-fourths reals during that time.[14] This sixty-year-old mestizo lived in Orizaba and appears in the census as a servant in the house of José Morales at Calle de la Campana, number 11.[15] The reasons for Corona's working on the Puente de Escamela for a full year are of course not stated in the records, but, according to the census manuscript, in November 1791 he continued to reside in Morales's house while he was in Domínguez's employ. He had no family of his own at the Morales residence. Corona's housing costs were probably relatively low, if indeed Morales charged him at all. Perhaps Corona was accruing the money to pay off a debt owed to Morales, or he may have even been able to use this income for something besides bare survival. Gregorio Corona clearly stands apart in the records from the typical day laborers at the Puente de Escamela. These latter peónes not only worked for a week or two and then usually moved on, but also frequently worked less than the full period of any particular workweek.

In any given week, approximately one day laborer out of each four hired to work on the bridge for a week left to do other things and thereby either lost or actually gave up the opportunity to earn some portion of the available wage. This pattern appears consistently throughout the eighteen months of construction work on the Puente de Escamela. Domínguez hired an overwhelming majority of these laborers again at some later time, so it seems unlikely that the sobrestante fired any of them for sloth or incompetence.

Some workers appear to have left for reasons perhaps not of their own choosing, such as twenty-eight-year-old Diego González. This man worked during each of four consecutive weeks in July and August 1791, but then never again appeared on a Puente de Escamela payroll. González did, however, appear in the Revillagigedo census report submitted in November 1791. He was in the Orizaba jail on a murder charge.[16] He obviously represents an extreme case, but many others also left the job never to return. The documents clearly show that Domínguez repeatedly rehired people who had quit on him previously. The laborers clearly exercised considerable flexibility in their periodic decisions not to work, which gives rise to another question: Did overseer Domínguez often have to cope with a shortage of available peón labor?

The reports for other weeks clearly suggest that Domínguez did not immediately replace laborers who left the job. During the week ending April 30, 1791, for example, eight men worked less than half the available workweek. Another fourteen worked less than the full week, but they did work more than half of it, so none of them could have been hired to replace another man who worked more than half the full workweek. Thus, the construction crew that began that week with a reported twenty-six day laborers finished it with no more than twenty, assuming that every laborer who could possibly have been hired as a midweek replacement was in fact hired as such.[17] Blas Francisco worked only one day during the week of May 8, 1791, but he could have replaced only one of the thirteen workers who did not finish out the week.[18] A similar analysis of the report for the week of October 30, 1791, shows that there could not have been any replacements hired for the nine out of sixteen laborers who worked less than the full, five-day workweek.

Although approximately one-fourth of the payroll entries were for less than the maximum workweek, Domínguez himself experienced much less than a 25 percent shortfall. Multiplying the length of each workweek by the number of laborers hired for that week and then comparing that result with the total number of paid man-days reveal a clearer indication of the magnitude of the sobrestante's problem. Domínguez paid for 8,011 man-days of work from the day laborers that he hired over the course of the project. That figure is 92.1 percent of the total 8,701 man-days available to him, assuming he never hired a midweek replacement day laborer and never extended the length of any workweek. These figures therefore equate to an absentee rate of 7.9 percent. One must remember that the laborers' tasks were in all probability not as precisely scheduled as for a modern-day assembly line, and the results, taken together, imply that Felipe Domínguez was generally able to get what he needed out of the men whom he hired in any given week.

Even so, Domínguez may at times have had some difficulty mobilizing the labor supply. At the beginning of the project, a few weeks were dedicated to quarrying stone and then hauling it down to the site. One Basilio Sánchez worked for three days during the week of March 30 and one day during the following week. Never again did Sánchez work on the Escamela project. The records also reveal that this same Basilio Sánchez was an arriero (muleteer) paid to haul stone from the quarry to the riverside.[19] The reason why Sánchez accepted work as a day laborer does not appear in the record, but the most logical explanation is that when he came to the quarry to haul stone, no full load was ready. In order to get his load more quickly and get paid to haul it, perhaps Sánchez had to help cut the stone. If Sánchez truly needed work as a day laborer, he might well have worked for Domínguez at some later time. Many other people did exactly that.

Any initial difficulties Domínguez had in attracting workers appear to have subsided after a few weeks, when it likely became more widely known that an opportunity to pick up a few cash reals had presented itself. The market for day laborers appears to have supplied the overseer with most of the man-days he wanted in any given week. This labor market must also be analyzed from the opposite side of the exchange: Were these day laborers able in some degree to manipulate that market to accommodate their own preferences and needs?

The Puente de Escamela records reveal the apparent paradox of intermittent work by persons who received low wages and might therefore be expected to work as often as possible. Perhaps Domínguez tried to spread the work among a flock of prospective work seekers each Monday, but that possibility leaves unexplained the repeated appearances of the same nine men, such as Gregorio Corona, on the payrolls. The turnover of names from week to week perhaps reflects personal choice on the part of some workers. We have seen evidence in both the 1767–70 records from Xalapa and the Puente de Escamela records that many agricultural workers appear to have hired on periodically to earn a few reals in hard cash.

Of course, the records reveal scant information about where else these persons may have found work, but obviously some of them might have juggled other work opportunities. At least one Puente de Escamela day laborer appears to have been a cigar maker at Orizaba's Royal Tobacco Factory. The 1791 census lists seventeen-year-old Francisco Aguilar's occupation as *purero*. This creole youth lived with his nineteen-year-old brother, who also rolled cigars, and their two sisters at Calle de Santa Rita, number 6. Aguilar worked a three-week stint as a day laborer at the bridge during January 1792.[20] Deans-Smith shows that these tobacco factory workers deftly manipulated the royal monopoly's piecework regime to their own advantage, including the practice of selling their *tarea* (daily production quota) to another person and then going off to find other work.[21] Without detailed weekly employment records from the Royal Tobacco Factory, one cannot determine if Francisco Aguilar was a regular employee who worked an angle or if he rolled cigars only on an intermittent basis and was passed over for employment that January. Either possibility is consistent with his appearances on the Puente de Escamela payrolls.

Felipe Domínguez was clearly willing to rehire day laborers who returned after weeks-long absences. The reappearance of so many names strongly suggests that the peónes at the Puente de Escamela were not forced to show up and sign up. Most of the short-stint day laborers who were identifiable in the 1791 census were agricultural workers, so their principal occupation was not that of manual laborer whenever and wherever the opportunity presented itself. If these peónes were indeed utterly desperate to get work at the Puente de Escamela, then it seems likely that more of their names would have reappeared more often on the weekly payroll reports, even if a crowd of hopeful men competed fiercely each week for a nod from overseer Domínguez. It thus appears that the unskilled labor market in Orizaba afforded rural workers some flexibility while still satisfying employers' needs.

Unskilled day labor was no longer the virtually exclusive province of indigenous persons. Members of all calidades enumerated in the Orizaba census appear frequently on the lists of the lowest-paid laborers. José Alberto, a twenty-year-old mestizo, was married to an yndia and is listed as an operario residing on the Rancho del Despenadero of don Agustín Rangel. José Alberto worked on the bridge for five and one-half days during the week of April 22, 1792, for which he was paid one peso and five and three-fourths reales. Juan Ramos, a thirty-year-old castizo also married to an yndia, was an operario on the tobacco rancho of doña María

González. Ramos earned three pesos and one real during the weeks of May 29 and June 5, 1791. Fifty-four-year-old Manuel Yslas also lived on the González rancho. A creole, this vaquero was married and had a daughter and three sons. He was employed as a day laborer on the bridge during the weeks of May 8 and May 15, 1791, where he also earned three pesos and one real.[22]

José Alberto and Juan Ramos had no children for which to provide, and their wives almost certainly worked on the tobacco ranchos as well. An assessment of their standard of living invites comparison with urban laborers based on wage-rate comparisons, but these two men worked primarily in agriculture. A vexing degree of complexity attends any attempt to discern all sources of income for such persons. Before 1770, a typical hacienda worker paid approximately one-fourth of his income for food, but that fraction doubled by 1800. Based on stable apparent wages, Van Young estimates that during the second half of the eighteenth century the ravages of inflation had inflicted a 25 percent reduction in real wages on Mexican agricultural laborers.[23] The wage rose 25 percent between the 1768 Xalapa figure of two reals daily and the two and one-half reals paid daily at the Puente de Escamela in 1791–92, but that cannot have made too much difference to day laborers from the ranchos who did construction work for only one or two weeks per year.

Day Laborers on the Toluca Road

Between 1793 and 1795, the viceregal government undertook a major effort to construct a road connecting Mexico City and Toluca, in the center of a grain-producing area lying some sixty-four kilometers west of the capital. Recall from chapter 2 that beginning at a point somewhere between the two cities, Captain don Diego García Conde's Brigada del Levante worked eastward to Mexico City and Captain don Manuel Agustín Mascaró's Brigada del Poniente worked westward to Toluca.[24] The payroll records from the Brigada del Poniente bring into sharp focus a picture of declining standards of living for its workers, clear evidence of a toughening labor market and increased economic stress.

Although the number of day laborers hired for this project was very large in comparison to the Xalapa roadwork or the Puente de Escamela construction, so many of the weekly reports are missing that as full an analysis as was done for Orizaba is impossible here. One unbroken series of records does exist, covering thirteen weeks from August to October 1794. This short series from the Toluca road is itself only a fragment, but it nevertheless contributes to a fuller understanding of the labor market in late-colonial Mexico when studied in the contextual light of the complete series from Orizaba. These documents show as many as 280 peónes working in the Brigada del Poniente, and a letter from García Conde indicates that in February 1794 he had 570 day laborers working at one time in the Brigada del Levante.[25]

By the autumn of 1794, however, the magnitude of the effort had somewhat diminished, and the two brigades were hiring similar numbers of day workers. I examined weekly records of the Brigada del Poniente for the period August 4 to October 31, 1794, to determine if they corroborated or contradicted the trends observed in documents from the earlier projects. By combining into one work record every possible name congruency, as was done with the other projects, I found that no less than 891 individuals worked as day laborers during the thirteen weeks under consideration. Of this number, 406 appeared during only one week, and another 80 appeared in only one stint of two consecutive weeks. Yet another 80 names appeared in each of only two weeks separated by a nonworking week on the part of the individual. These three categories account for 63.5 percent of the total names, a figure somewhat lower than the 75 percent for the Puente de Escamela workers.

Among the Toluca workers, 53 (5.9 percent) appear in eight or more of the thirteen weeks under consideration, and even this low figure may be higher than the reality. For example, the data points for the name José Antonio were combined to yield one person who worked in each of the thirteen weeks. This name may indicate two or more actual individuals, so the 5.9 percent figure is in reality a maximum possible result. Because that percentage is quite low, it is clear that the Toluca day laborers exhibited a tendency for intermittent work on the roads, a tendency similar to that observed in their counterparts at Xalapa and Orizaba.

The Toluca road day laborers probably also worked primarily in agriculture. The records do not specify the precise location of the particular construction sites between Mexico City and Toluca, but presumably the Brigada del Poniente worked in the area to the east of Toluca itself. The Toluca area produced wheat and maize, much of it destined for the viceregal capital.[26] As the day laborers most likely hailed from the Toluca area, their lives would have been regulated by a different agricultural rhythm than in a tobacco area. Rural laborers in grain areas typically worked approximately eight months out of every year.[27] However, because day laborers from both tobacco-producing Orizaba and grain-producing Toluca show such similar patterns of employment on the road and the bridge, differences in the demands of agriculture seem to have exerted but little impact on their propensity to seek construction work. Considerations of other explanations must at this point be deferred until the discussion of wages later in this chapter.

A major difference from the earlier projects was the impressment of prisoners and debtors by the subdelegate of Toluca. This official expressed considerable personal interest in the construction of the road from his official seat to the capital of New Spain. In a letter to the viceroy, he argued that two birds could be killed with one stone if the vagrants of Toluca were put to work on road construction. The labor would improve their character, help them support their families, who were just then a burden on the sources of charity, and get them started paying off their debts.[28] The official made no mention of the creditors' identities, but one naturally suspects that the subdelegate himself must have been a prominent one. Captain Mascaró investigated the possibility of mobilizing this additional labor supply, and he must have been favorably impressed because he waxed enthusiastic in his own letter to Revillagigedo.[29] The viceroy's approval of an employment scheme appears to have followed shortly because Toluca fielded a construction unit of its own that submitted separate reports as a subordinate component of the Brigada del Poniente. *Peónes voluntarios* (voluntary day laborers) received one and one-half reals per day. The subdelegate also sent prisoners to work for one real daily, the same as that paid to the yndio repartimiento laborers of Xalapa in 1757. Given the tone and content of the subdelegate's original proposal, one suspects that a large portion of the wages reported as paid to these workers in fact ended up in the subdelegate's pocket. During August 1794, this Cuadrilla de la Ciudad de Toluca fielded first twenty-three "volunteer" day laborers and eleven prisoners, which numbers rose two weeks later to forty and fourteen, respectively.[30] Of particular note is that twenty-one of the initial twenty-three day laborers worked three consecutive full workweeks and at the lowered wage, a pattern completely different than in the main brigades or in the Xalapa and Orizaba projects. One can only conclude that the subdelegate of Toluca somehow coerced the volunteers into working on the road.

Both Toluca road-construction brigades made extensive use of child labor, a practice not evident in the Xalapa and Orizaba records. *Cabritos* were youngsters hired for tasks appropriate to their age.[31] They appear on the Brigada del Poniente records at wage rates ranging from one-half real to one and one-half reals per day. Initially, one group of cabritos received one and one-half reals per day as opposed to one real for the others, who presumably were smaller and only able to perform less physically demanding work.

The cabritos showed a very strong propensity for single appearances on the payroll records. Of 327 pay entries for cabritos, 209 (63.9 percent) appeared but once. Another 40 appear for only two consecutive weeks and then vanish completely. Thus, 76.1 percent of them put in single appearances, consonant with the 75 percent appearance rate shown by the Orizaba day laborers and the 63.5 percent rate shown by the peónes on the Toluca road. Only 8 cabritos (2.4 percent) worked in eight or more weeks, a fraction less than the 5.9 percent figure for day laborers, but both rates are still quite low. On balance, the young cabritos exhibited work habits generally consistent with those of older unskilled laborers.

Some of the cabritos may have been the children, or other relatives, of day laborers. In 76 day-laborer pay entries, there appears a name that exactly matches that of a cabrito for that week. For example, Joaquín Nava worked as a day laborer from August 4 to August 29, 1794. That name also appears on the list of cabritos for each of the four workweeks included in that period. In such cases, perhaps a youngster's wage was paid to an elder relative whose name appears on the record; the record's purpose was likely more to note for the Real Hacienda the disbursement of the Crown's funds than to identify to posterity the person who actually did the work.

But the record also suggests that some individuals may have been hired as day laborers in some weeks and as cabritos in others. Forty-five cabrito names match day-laborer names, but appear in different weeks than do their namesakes. Some of the name congruencies are doubtless coincidental, and in such instances there would actually have been two different persons with the same name. Consider the example of an individual named José George (spelled as given) who worked as a cabrito during the week of August 11 to August 16. The name is so unusual that it almost certainly refers to just one person. He appears in the following week as a peón, but then disappears completely from the records. Day laborers such as José George may occasionally have felt compelled to accept employment at the lower wage rate for child workers, which suggests that the man needed to work for whatever wage he could get for his labor.

Looking at yet another case, we find that José Yglesias worked as a day laborer from August 4 to August 16 and then as a cabrito for the week of August 25, then immediately again as a peón for a two-week stint from September 1 to September 13. He did not work at all during the week immediately following, but did reappear one last time during the week of September 22. Yglesias appears nowhere on the smaller lists of day laborers for the weeks of October 1794 and so seems to have gone elsewhere in search of work.

Although lower than the wage of a peón, the one real daily wage of a cabrito is still on a par with Taylor's findings for Oaxaca agricultural workers as noted earlier. Road work, however, offered none of the non-monetary sources of income that a hacienda might have, such as shelter and some food. If adult workers felt compelled to accept work at the cabrito wage, then they must not have been afforded much flexibility to limit the quantity of labor they chose to supply to the labor market. José George and José Yglesias probably worked as cabritos because they had been passed over during the selection of the day-laborer crew, but still needed any job they could get.

Each foreman could very well have favored a few individual day laborers that he knew well by hiring them each time he saw them among those gathered at the site seeking work, which might explain the few cases of frequently appearing peónes that each set of records contains. The foreman might then have rounded out his crew by choosing other men more or less at random, and the possibility that he took kickbacks from those that he chose certainly exists. Whether a disappointed laborer returned in later weeks or went elsewhere in search of work would depend, of course, on a variety of factors, and some of the workers with single appearances may not have been wanted back by the foremen after one appearance.

Because the Toluca road payroll records do not survive in a continuous series of more than thirteen consecutive weeks, finding examples of laborers who stopped working until depleting an accumulation of previously earned wages becomes virtually impossible. However, the Toluca road day laborers exhibited the same marked tendency for single appearances on short stints as did their counterparts at the Puente de Escamela a few years earlier and as did those working around Xalapa a few decades earlier. Deborah Kanter finds that haciendas in the central and southern portions of the Toluca Valley seldom relied on coercion to meet their labor needs, which suggests the existence of a large labor supply and a buyers' market. She also notes that many Nahua villagers had to leave their pueblos to supplement their income.[32] Most likely, many of the peónes and cabritos of the Brigada del Poniente worked intermittently on the Toluca road for that same reason. The fact that some men worked on the Toluca road as both day laborers and as cabritos, however, clearly implies that their economic situation may not have permitted them the luxury of limiting their individual labor supplies. In this respect, they seem to have sought whatever work they could get for practically any wage. In search of more evidence and yet another vantage point from which to evaluate trends in the late-colonial labor market, we can turn to an examination of wages themselves.

THE WAGES OF UNSKILLED LABOR

When considering the matter of just how well off (or poor) people in any society may have been, we must look at two obvious questions: How much money did they earn, and how far did that money go? Wage series go a long way in assessing the economic status of the persons who work for wages, although we must constantly keep in mind that the availability of work and the laborers' employment habits are other factors in the equation.

As we have seen, the transition from native village draft labor to free-wage labor between 1757 and 1767 was attended by a rise in the daily wage from one to two reals. This increase can be seen as a response to resistance to the draft-labor regime, such as that which gave rise to the complaints of the alcalde mayor of Xalapa, don Antonio Primo de Rivera. At the Puente de Escamela in 1791 and 1792, day laborers commanded a wage of two and one-half reals per day, a wage consistent with the daily wage scales for employees of the Royal Tobacco Factory and for construction laborers in Mexico City.[33]

Deans-Smith shows that debt peonage persisted in Orizaba tobacco ranchos even after 1800. Consistent with Van Young's finding that enganches were not generally used to recruit temporary workers, however, the Puente de Escamela records contain no evidence of an incentive payment to any worker at any time during the construction of the bridge.[34] In fact, wages were temporarily reduced during the week of August 14, 1791. For each of two days out of a workweek of five and one-half days, every day laborer and stonemason suffered a one-half real pay cut. Only overseer Domínguez's daily wage of six reals remained untouched. The record shows no reason for the reduction, nor does it contain any evidence of subsequent restitution, in cash or in time. Thirty-two of the forty workers who received reduced wages returned to work at the bridge the following week, a rate of reappearance typical of other weeks.[35] Wages remained generally fixed at their earlier levels for the duration of the project, although, as discussed in detail in chapter 5, a few day laborers were hired as stonemasons during the last few weeks of construction. These few persons received a lower wage than that which had been paid to stonemasons earlier on.

The analysis has thus far considered only the matter of "apparent" as opposed to "real" wages and so has taken no account of the effects of inflation on the purchasing power of the money earned by a person on one of the construction projects. Their apparent wage having risen from two to two and one-half reals per day between 1767 and 1791, day laborers appear to have more or less held their own. This 25 percent increase in the daily apparent wage would keep their real wage at roughly the same level, based on Van Young's finding that persons with a stable wage suffered a 25 percent erosion of their ability to purchase food by the end of the eighteenth century.[36] By 1794, however, they had lost that ground.

Day laborers in both brigades on the Toluca road received two reals per day. Particularly noteworthy is that the Toluca workers received a wage 20 percent lower than that of the Puente de Escamela day laborers. That decrease exactly matches the wage reduction suffered by Mexico City construction day laborers between 1790 and 1794.[37] Urban workers would have been even harder hit by the inflationary pressure on real wages than were agricultural laborers. The absence of a consumer price index for Mexico in the 1790s makes a mathematical estimate virtually impossible, but any inflation had serious effects on the poor in a preindustrial society.[38]

The Mexico City construction workers appear to have suffered their wage reduction some time in late 1793 or in 1794. A report of expenses by the Royal Tobacco Factory for the week of May 27 to June 1, 1793, indicates that 220 day laborers worked for six days, together earning 404 pesos.[39] This equates to a daily wage of two and one-half reals, the same that was paid at the Puente de Escamela in 1791 and 1792. In Mexico City, the daily wage for construction labor had decreased to two reals by 1794. The unskilled laborers' position had worsened, and for some of the Toluca road workers even more trouble lay just ahead.

The wage situation of the day laborers in the Brigada del Poniente took an additional turn for the worse when they found their daily wage reduced from two reals to one and one-half reals during the week of September 9 to September 14, 1794. On that week's pay record, 116 day laborers appear, down from the previous week's 234. Perhaps a large number of workers walked off the job after receiving word of the pay cut, but 212 day laborers were at work during the immediately subsequent week of September 15 to September 20. This total includes 61 peónes who also worked during the week of the cut, a turnover rate that is typical of all the records of the Brigada del Poniente. Twenty-five names disappear completely from pay records after September 13, so perhaps these individuals did in fact refuse to return to work on the Toluca road subsequent to the 25 percent wage reduction.

The day-laborer workforce thus found itself unable to resist a major wage cut. Unskilled peónes could be replaced easily, so they lacked the collective clout enjoyed by more skilled groups such as the silver miners of Real del Monte, who launched a successful strike in 1766. One historian analyzes the miners' successes and finds that those workers exercised a cohesive class consciousness in their resistance to reductions in wages and perquisites.[40] Skilled miners did not exist in great supply, however, even with the increasing population of late-eighteenth-century Mexico. A second scholar has shown that labor was always in short supply in the colonial Mexican mining sector, and still others contend that the supply remained small enough that miners "constituted a labor aristocracy."[41]

Unskilled peónes did not enjoy such an advantageous position in the labor market. A day laborer in construction could be replaced much more easily than could a skilled pick man from a silver mine. Here we see a change from the situation of a century earlier. A study of hacienda labor in seventeenth-century Tehuantepec shows that persons in the free-wage sector did not show up for work if the hacendado or *arrendatario* (tenant farmer) failed to pay wages in a timely fashion. In that case, the local population was in numerical decline, a situation opposite to that which existed in Mexico in the 1790s.[42] The toughening labor market of the late Bourbon period compelled the day laborers of the Brigada del Poniente to return to work and endure the reduction of their daily wage to a mere one and one-half real.

The Brigada del Poniente wage cut stayed in effect throughout the remaining weeks covered by the documents analyzed here, but two individuals' wages were restored to two reals per day. These men appear under the names Chávez and Victoriano. Each of these two men worked all thirteen weeks of the period, and they are singled out and identified as *capitanes* on the lists of day laborers who received the one and one-half real wage. The role Chávez and Victoriano may have played in inducing other day laborers to continue working for lower wages remains a matter for speculation. It seems more likely that pressures of the labor market would have greatly outweighed the importance of any blandishments by Chávez or Victoriano, especially because the reduced numbers of day laborers' names that appear in the subsequent weekly reports suggests that the opportunities to find work building the Toluca road had become rather more scarce.

The day laborers of the Brigada del Levante continued to receive two reals per day throughout the period, and the surviving documents contain no evidence that wages were ever reduced there. The disparity in wages suggests that the decision to reduce the wages did not originate at the viceregal level, and that the authority to make the change rested no higher than Mascaró and García Conde, the military engineers in charge of the brigades. Unlike the documents from the Xalapa projects and the Puente de Escamela, the Toluca road payrolls do not indicate the exact location of a specific construction site, which precludes the detection of any relationship between differing locales and individual wage rates. The Brigada del Poniente documents also give no indication that Mascaró's action had any effect on García Conde's workers. The wage rates for day workers on these public-works projects thus could not have been set by viceregal fiat, even if the viceroy's permission was obtained, and the differences between the two Toluca road-construction brigades suggests that central Mexico's unskilled labor market was by no means fully integrated.

Recall from the preceding chapter that a shortage of funds led to cost-cutting measures on the Toluca road-construction project. This was probably the immediate cause of Captain Mascaró's reducing the peónes' wage to one and one-half reals, but obviously the labor market was able to bear the blow because the work continued apace. The Toluca subdelegate's reports from August 1794 indicate that he paid peónes voluntarios one and one-half reals, whereas his prisoners received only one real per day.[43]

Records from construction of the Camino de Luisa do not survive, but it seems a quite reasonable assumption that they would tell us the same things as the records from the Puente de Escamela and the Toluca road. Repairs also continued to be a major undertaking. Diego García Conde, by then promoted to lieutenant colonel, filed a report in February 1804 indicating that a large number of persons were engaged in repairs to the Camino Real around Xalapa. Out of a total 1,175 individuals, 385 appear to have been prisoners quarrying stone. Another 365 were divided among four crews working east of Xalapa at Cerro Gordo. At San Miguel del Soldado, 191 free-wage laborers worked, many of whom doubtless came from the same village that twenty-three years earlier had been dispossessed of its maize field in order to create the extant roadbed.[44] The work continued.

Artisan Road Laborers

During the eighteen-month period in 1791–92 covered by Felipe Domínguez's payroll reports for Orizaba's Puente de Escamela, twenty-four individuals were hired as stonemasons. These tradesmen earned five reals per day, twice what was paid to a day laborer. Like those in the 1767–68 documents from Xalapa, these stonemasons also exhibited work patterns that differed greatly from those of the day laborers. Only five of the twenty-four worked for short stints of one or two weeks, whereas the others worked for periods averaging eighteen weeks in length, with but rare interruptions never lasting more than one full workweek. Two stonemasons, José Marcos and José Marcelino, returned to work at the bridge in October and December 1791 following absences of fifteen and seventeen weeks, respectively.[45] These two men's protracted absences from the project may well have been determined by an agricultural cycle, such as the timing of a harvest. Two other stonemasons, Miguel Casimiro and Miguel Barbosa, returned to bridge construction after interruptions of six weeks and two weeks, respectively. Excepting José Secundino, who never returned after the August 1791 temporary wage reduction, the rest of the stonemasons worked each week for long stints.[46]

Five of the twenty-four stonemasons were men who had earlier worked at the bridge for significant periods as day laborers. Felipe de Santiago worked as a stonemason for three weeks during late November and early December 1791 and then returned in January 1792 as a day laborer.[47] While working as a peón, Santiago had returned to work at Escamela as many as seven different times. After working fifty weeks as a day laborer, Juan Estevan worked as an albañil for the final five weeks after the principal stonemasons apparently moved on. Domínguez's records indicate that no major structural work was undertaken, the tasks during that time being limited to relatively simple jobs such as surface facing and ornamentation. For the last three of the five weeks, Juan Estevan received a daily wage of four reals in lieu of the normal five. Possibly Alcalde Ordinario don Gregorio Bezares took overseer Felipe Domínguez to task over whether Juan Estevan was truly an albañil worth the going rate and directed a wage reduction, or perhaps the required tasks were of insufficient complexity in the alcalde's mind to justify paying the going rate for an albañil.

Working for longer stints and at a higher wage, the stonemasons earned considerably more than the typical day laborer did on road work. The reasons for their comparatively "Calvinistic" work habits are unclear, but one strong possibility is the simple fact that they were getting better pay than the unskilled day laborers.

Only two of the twenty-four stonemasons appear in the Revillagigedo census of 1791. Of the 301 day laborers, 42 can be confidently correlated with the census record, a rate almost double that of the higher-paid stonemasons. The workers who do not appear in the census may have come from other locales in search of work, or they may have been Nahua from the Orizaba area. Yndios were ineligible for military service, and as tributaries they paid taxes under a different jurisdiction compared to that of members of the república de españoles. Because the purposes of the census were to enumerate persons for possible conscription and to establish up-to-date tax rolls, yndios were not identified by name in the 1791 census.[48] Future research may fill in the large gaps resulting from leaving the indigenous people nameless, but it seems likely that these stonemasons were Nahua. Recall that the stonemasons on the Xalapa projects appeared to be overwhelmingly Totonac, based on a tentative comparison of the 1767 overseer reports with the 1777 census. A similar comparison of that census manuscript with the Escamela records yielded eleven name congruencies among the twenty-six stonemasons. Three were mestizos, five were yndios, and three others had matches from both the mestizo and yndio lists. The less conclusive results of the Orizaba comparison may mean that the composition of the villa's stonemasons had always been more diverse, even though Orizaba and Xalapa are often considered to have been places with quite similar societies.[49] Conversely and rather more likely, it may mean that stonemasonry was no longer so strongly considered "Indian work."

Manuel Estevan worked 258.5 days as a stonemason at Escamela between May 5, 1791, and June 23, 1792, earning a total of 161 pesos and 3.5 reals. A mestizo, Estevan resided in Orizaba's Barrio de Ixhuatlan with his wife and son. His age is not recorded, but his occupation is listed as operario.[50] His earnings prorate to 138 pesos per year, which is within the band of 96.5 to 196.5 pesos per year, three-fourths of Haslip-Viera's estimated cost of supporting an average working-class family of 3.8 persons in Mexico City.[51]

The other stonemason who appears in the Orizaba census is Pedro Nolasco, a fifty-five-year-old español who lived in Orizaba with his mestiza wife, Juana Victoria. The census indicates that no other family members resided with them, so either their children were grown and had left home, or, somewhat more likely, they were childless or had no surviving children. Pedro Nolasco worked as a day laborer at Escamela for 18.5 days between July 26 and August 27, 1791. He remained on the job after the August 14 wage cut and returned to work the full week immediately following the reduction. He then left the project, not returning until November 13, 1791, at which time he began a long stint as a stonemason that lasted until May 5, 1792.

In the ten months that Pedro Nolasco worked on the Puente de Escamela, he earned 73 pesos and 6.25 reals. This is proportional to an annual earning of about 90 pesos, well above the previously estimated minimum threshold for a working-class family of two. Even if Nolasco had found no other work for the two months that remained until the anniversary of his first workday at the bridge, his earnings at Escamela would still have sufficed for him and his wife for that year.

Most of the Escamela stonemasons earned enough at the bridge to sustain an average working-class family. The two for whom the census provides data regarding family size were probably able to make ends meet on the one income, although no account is taken here of taxes, tithes, or tribute. Future research may reveal whether such artisans could reasonably expect to find sufficient work over the long haul to maintain their position. As a group, however, they appear to have done considerably better than did the unskilled day laborers during the eighteen-month construction period at Escamela.

At five reals per day, the Orizaba stonemasons received double the wage of the day laborers. Recall that in 1767–68, the Xalapa albañiles received four reals, a figure that was also twice the daily wage of the unskilled laborers on those projects. Applying to the stonemasons' situation Van Young's estimated cost-of-living increases over the second half of the eighteenth century, one would conclude that, like the day laborers, the stonemasons held their own or perhaps even did somewhat better because their wage rose proportionally, whereas increased food costs would have affected both groups equally in absolute terms. However, not all artisans seem to have fared as well. Carpenters appear on the various Xalapa records, but none is recorded anywhere on the Puente de Escamela or Toluca projects. Because the nature of the work was identical, one wonders what happened to these tradesmen.

Linking the Puente de Escamela documents with the 1791 Orizaba census manuscript shows that at least one carpenter was in fact hired to work on the bridge. A resident of the villa of Orizaba, José Mariano Navarro was married and had three sons.[52] The records indicate that his position was clearly deteriorating. This thirty-one-year-old mestizo carpenter worked on the bridge for 47.5 days between August 28 and October 23, 1791, earning in the process a total of 112 pesos and 6.75 reals at the standard rate for day laborers of 2.5 reals per day.[53] As previously noted, carpenters on the 1757 and 1767–68 projects around Xalapa commanded 3 reals per day.

Why would Navarro work for less? He possibly hired on as a day laborer to fill in a slack period, but this seems unlikely. Other stonemasonry projects had needed carpenters, and the Puente de Escamela was constructed with well-tried methods that would have included the use of scaffolding and other ancillary carpentry. The records of purchase of construction materials substantiate the hypothesis that Navarro was actually hired to work as a carpenter. The same documents that contain the payroll information also show that throughout the project, Domínguez frequently received shipments of a variety of construction materials, including stone, sand, lime, iron, and wood. Only once did he receive a shipment consisting exclusively of wood. That shipment was paid for on September 18, 1791, in the middle of the period during which the carpenter Navarro appears as a day laborer on Domínguez's payroll.[54] The arrival of this unique shipment suggests that a greater-than-normal consumption of wood occurred at that point in time and that the wood was being used in carpentry projects. After several weeks of very steady employment at Escamela, Navarro's name vanished permanently from the payrolls. Domínguez very likely hired Navarro for his carpentry skills and not merely to provide one more strong back for unskilled manual work.

Navarro probably left to find work elsewhere in his trade. He must have reasonably expected to be able to find it, and perhaps he did not feel forced to continue work as a day laborer in order to make ends meet. Nevertheless, the clear fact remains that this presumably skilled artisan earned one-half real less per day than he would have received twenty-three years earlier in nearby Xalapa. It is unclear if a carpenters' guild (*gremio*) existed in Orizaba in 1791, but the records do indicate that there was one by 1803. In either case, carpenters themselves appear to have been unable to sustain their position in the labor market, and a 16.6 percent reduction in apparent wages must surely have translated into a substantially reduced living standard.[55] Haslip-Viera finds that carpenters employed by the Royal Tobacco Factory in Mexico City received six reals per day in 1776, but that in 1794 their wage had declined to four or five reals. The Mexico City carpenters still appeared as a separate category of artisans who received more money than did day laborers. This was also the case in the Brigada del Poniente building the Toluca road, where one carpenter regularly received three reals per day instead of a day laborer's two, and another received four and one-half reals during the one particular week that he worked for Captain Mascaró.[56] Navarro's case shows a more precipitous decline than that experienced elsewhere, but his experience still parallels the general decline in economic status that these artisans experienced in Mexico during the later part of the eighteenth century.

The magnitude of the estimated deterioration of Navarro's economic position increases if account is taken of inflation. Not only did Navarro earn a lower wage, but its value as a real wage declined even more. Richard

Garner tentatively estimates an inflation rate of one-half percent per year during the eighteenth century, with a somewhat higher rate during the second half of the century.[57]

The wage schemes for albañiles building the Mexico City–Toluca road differed between the two construction brigades. The report for Captain don Diego García Conde's Brigada del Levante for the week beginning September 1, 1794, shows twenty stonemasons at five reals per day, three more at six reals, and one who received one peso and four reals per day.[58] In Captain don Manuel Agustín Mascaró's Brigada del Poniente, however, most of the stonemasons earned either four, four and one-half, or five reals per day. Two albañiles earned six reals, and the leading stonemason earned one peso per day.[59] Although their average base wage was lower than in the Brigada del Levante, the records show that the stonemasons of the Brigada del Poniente also received an incentive payment of forty reals for each twelve varas of paved surface that they completed. Recall that Mascaró was the officer who reduced the day laborers' wage by 25 percent.

The stonemason section of each Brigada del Poniente payroll records a payment for completed distance but offers no hint as to how the bonus was divided among the individual workers. Entry of the payment in the stonemasons' section of the report implies that members of no other group shared in the spoils. If the money was divided equally among them, then each stonemason usually received an additional twenty to twenty-five reals per week. More likely, the higher-paid albañiles got a larger fraction of the money, but in either event each artisan probably would have netted an extra 50 to 60 percent of his base wage. Three of the thirteen reports, however, list extraordinary figures that suggest that a particular few stonemasons manipulated the incentive payment to the detriment of others.

The report for the week of September 1 to September 6, 1794, indicates that only six stonemasons completed 4,100 varas during that week, as opposed to the preceding week in which eighteen workers are credited with completing 1,320 varas. The six individuals in the former case would each have taken away an extra twenty-five pesos and five reals. In another suspicious case, four stonemasons averaged about ten pesos and two reals each for finishing 1,100 varas of pavement during the week of October 13 to October 18, immediately following a week in which sixteen stonemasons received credit for only 360 varas.[60]

Perhaps circumstances such as adverse weather so greatly prevented progress in some weeks that the reports are in fact accurate from week to week, but the reports for the last two weeks of October 1794 show small numbers of stonemasons completing much longer than average stretches of pavement and receiving an extra five to ten pesos in those weeks. Several names repeatedly appear in these shorter weekly lists, and it seems likely that a few stonemasons seized the opportunity to take advantage of the other stonemasons by keeping them off the payroll when a large bonus was forthcoming. Project officials must have lent at least their tacit approval, if not their outright connivance, and their own pocketbooks perhaps benefited as well.

The stonemasons of the Brigada del Poniente displayed a proclivity for intermittent work habits that rather exceeded the habits observed by albañiles in the other projects. Forty-three names appear in the thirteen selected pay records from the Toluca road documents. Sixteen of those names are single appearances of only one week, and three more are single appearances of two consecutive weeks, thus accounting for 44 percent of the aggregate number of stonemasons. At the Puente de Escamela, just four (16.6 percent) of twenty-four stonemasons had only single appearances. Not only is the gross fraction much lower, but the Orizaba records cover a full eighteen months and reveal no instances in which a stonemason returned to that project after an extended absence. Eight of the Toluca stonemasons (18.6 percent) worked at least eight of the sampled thirteen weeks, whereas only two of the Orizaba men worked for the same fraction of the total time indicated in those records. This result is inconclusive in comparing the tendencies of the two groups to work long stretches, however, because the Orizaba percentage could be made to exceed that of Toluca merely by selecting a thirteen-week sample between May and October 1791.

An additional complicating factor stems from the possible manipulation of the Brigada del Poniente stonemason workforce discussed earlier, when several frequently appearing stonemasons may have been refused the opportunity to work in weeks during which a large bonus for completed pavement was planned to be awarded. Even so, the Toluca stonemasons show a much higher propensity to return to work on the road than did the day laborers. Here again emerges the paradox that those who seem to need the work most desperately also appear less likely to stay on the job, the same case as found in the Xalapa and Orizaba records. But recall that the free-wage day laborers most likely were rural workers who intended to work only a short stint and return to their crops.

CONSIDERATIONS OF THE LABOR MARKET IN MEXICO

The transition from repartimiento village drafts to free-wage, unskilled day laborers was not achieved all at once. Viceroy Antonio María de Bucareli indeed chided the Ayuntamiento de Puebla in 1776 for its request that he authorize it to use forced labor to repair a key bridge, but as Doris Ladd shows, this same viceroy had authorized the Conde de Regla to use press gangs to round up laborers for his silver mines. Sergio Florescano Mayet notes the objections raised by hacendados when the possibility of returning to a village draft-labor scheme arose, from fear of diversion of their workforce to road labor. Brian Hamnett finds that the viceregal government often served as a sort of mediator between various employers and the labor force and that colonial authorities consistently preferred free labor.[61] Fits and starts aside, the trend toward free-wage labor for public-works projects in eighteenth-century central Mexico seems a clear one.

An increasing reliance on free-wage labor suggests an expanding labor supply, as demonstrated in the works of several historians studying different regions of Mexico during the period considered here.[62] It also represents a turn away from a practice that had been customary for more than two centuries, which in turn suggests another question: Was colonial New Spain a "modernizing" society or a "traditional" one? The answer obviously cannot be that it was either strictly "modernizing" or "traditional," and the tension between the two conceptual extremes demonstrates the simultaneous importance of each. Custom retained its importance, but stasis did not exist. Changes in wage practices further illustrate the point.

In the 1767–70 Xalapa projects and the Puente de Escamela project in 1791–92, artisan albañiles received a wage exactly double that of unskilled peónes, and these wages equaled the going rates for similar workers in Mexico City.[63] In his classic study of the English working class, E. P. Thompson contends that at the beginning of the nineteenth century custom rather than supply and demand still determined an artisan's wages. He also finds a persistent but declining pattern of payments in-kind to English rural laborers.[64] Great Britain undoubtedly stood at the forefront of processes of industrialization and "modernization" in the eighteenth-century Atlantic world, but a traditional system of wages is still held to have predominated.

Although historians have documented the wage-cutting efforts of the silver mine owners, wages in Mexico also seem to have followed customary practices during the eighteenth century. Matthew Restall finds that in Yucatán, Afro-Mexican and Maya workers typically earned one and one-half reals per day.[65] The evidence considered here shows that adherence to customary practice continued to be common in the 1790s, but from the Toluca road-construction records emerges evidence of some significant changes: 25 percent wage cuts for unskilled day laborers, and a lowered base wage for stonemasons, coupled with a schedule of productivity incentives. These "modernizing" changes were used only in the Brigada del Poniente, whereas the Brigada del Levante workers continued to receive wages according to a traditional system.

In the context of a traditional wage, the general reduction of day laborers' wages between 1791 and 1794 also shows the operation of "modernizing" forces. The cut from two and one-half reals to two reals per day must certainly have reduced the economic position of the peónes to lower than that to which they had been accustomed, especially taking inflation into account. The day laborers of the Brigada del Poniente had little alternative but to endure a yet additional reduction to one and one-half reals. In other words, these workers could take the wage cut or leave it. Forces of supply and demand shaped the colonial Mexican labor market to no small degree.[66]

Eighteenth-century inflation rates of approximately 1 percent may appear insignificant to the modern reader, but as Garner notes, "[e]ven modest inflation in a pre-industrial economy could have a disruptive effect on the consuming public."[67] Also, recall once again Van Young's estimate that food costs roughly doubled between 1750 and 1800. If inflation between 1768 and 1791 is assumed to be 1 percent per annum, then an overall 26 percent increase in the cost of living occurred. If the inflation rate is assumed to have been only one-half percent per year, then the total increase in the cost of living would have been 12 percent. In 1768, the Xalapa stonemasons earned four reals per day. In 1791, those working on the Puente de Escamela received five reals for each day on the job. Much if not all of that apparent rise must have been offset by inflation.

We have seen that a day laborer earned twice the cash of a hacienda peón, but that the latter received substantial payments in-kind. We have also seen that stonemasons and other tradesmen often earned twice as much as did the unskilled laborers. To place the analysis of late-colonial Mexican wages in a broader context, it is useful to consider the incomes of some other persons whose lives also connected to the Camino Real.

A few yards away from Orizaba's Puente de Escamela stood a customs gate. At the Garita de Escamela, *guardas de la renta* of the Real Aduana collected road tolls and duties on goods from the Camino Real traffic. Most of these customs guards received 365 pesos annually, which prorates to 60 percent more than a stonemason's daily wage. Three of the Escamela guards were hired after 1780, and these men received only 228 pesos and one real per year, the same daily rate as paid to the albañiles working on the nearby bridge. The records do not indicate why the pay for Orizaba-based customs guards was reduced, whereas those who were hired at the same time in the city of Puebla continued to receive a full peso per day.[68] The Orizaba guards who had been hired earlier continued to receive the higher rate, but before jumping to equate the other three with the stonemasons, we should note that customs guards were paid for each day of the year, whereas the tradesmen had to go out and find work for each day's pay.

Orizaba was a tobacco-producing center whose output moved along the Camino Real. The permanent staff of the villa's Royal Tobacco Factory included a sizeable guard force that prevented theft, maintained order in the workplace, and ensured that tobacco growing and cigar fabrication were conducted in accordance with the rules of the Bourbon regime. The tobacco guards made an annual salary equating to twelve reals per day, considerably more than the customs guards, but the latter were quartered at the garitas. The 1791 census manuscript shows the residences of the tobacco guards scattered throughout the town, so presumably they had to procure their own living quarters.[69] Don Antonio Sobrevilla, the Iberian-born chief of the Orizaba tobacco guard force, earned 1,500 pesos per year, almost triple the income of his guards. This salary allowed Sobrevilla and his extended family to live comfortably among the planter elite on the Calle de las Damas near the center of town.[70]

Major construction projects such as the Toluca road and the Camino de Luisa were carried out under the direct supervision of military engineers. In that day and age, these technically specialized army officers were foremost in the Western world in the requisite skills and education. Captains Mascaró and García Conde were two of the ten such officers assigned to the Viceroyalty of New Spain in 1795. As captains of infantry in the regular army, they received annual salaries of 912 pesos plus rations in-kind.[71] Their additional status as *yngenieros ordinarios* raised their cash pay to 2,500 pesos.[72] In contrast, had a day laborer in Mascaró's Brigada del Poniente been inclined or even able to work six days in each week of the year, he would have earned only 78 pesos *before* the 25 percent wage cut. Quite a difference!

At this point, it is well to rehabilitate the captain. Just who was this man who cut the workers' wages? Born in 1747, Manuel Agustín Mascaró was a native of Barcelona who in 1764 became a military cadet. After receiving the king's commission, he campaigned in Algeria and then served as instructor of mathematics at the Real Academia in Madrid from 1774 to 1776. He was ordered to New Spain in 1778, where he apparently spent the rest of his life. Besides the Toluca road, he developed plans for extensive modernization of the fortifications around Acapulco, planned and supervised the construction of the important Arispe Dam near Mexico City, and completed numerous other bridges, buildings, streets, and roadways. He was particularly renowned for his map-making skills.[73]

Mexico in the nineteenth century suffered from a lack of physical infrastructure, as the Bourbon regime well knew for years. Much of what did exist, however, resulted directly from the efforts and talents of the few military engineers on duty there during the final decades of Spanish colonial rule. Structures engineered by the likes of Mascaró and García Conde won the unqualified admiration of not only travelers, but also professional observers such as the West Point–trained engineers serving in the U.S. invasion forces of 1847.[74]

Captain Mascaró mostly certainly must have been a tough boss. But before singling him out for special condemnation for using oppressive measures, modern-day observers should temper their opinions by first taking account of the context of the times. Whether vilified in Mexican historical memory as is the Marquis of Branciforte or lionized as are the Marquis of Croix and Count Revillagigedo the younger, the Bourbon viceroys were regalist dictators engaged in the intensification of the political and social control of Mexican society. That they genuinely meant this control to have beneficial results matters not at all to the fundamental nature of the system itself. An active agent of colonialism, Mascaró nonetheless remained one part of a larger system. In other words, he made his own history and forced others to make theirs, but under conditions none of them chose.

Summary

Taken together, the four sets of road- and bridge-construction expense records analyzed in this chapter and in chapter 3 offer many insights into the labor market of late-colonial Mexico. Demographic conditions supported an increased use of free-wage labor, but in 1798 Viceroy Miguel José de Azanza reported to the Crown that a labor shortage continued to exist in Orizaba, Xalapa, and Veracruz.[75] Such conditions presage some form of coercion to rectify the shortage, and at times the specter of draft labor reared its head. The inspector of Orizaba's Royal Tobacco Factory proposed in 1805 to assign some Nahua pueblos and barrios to tobacco cultivation and others to road work. Shortly after the drafting of that plan, an official of the Real Tribunal del Consulado attributed a shortage of tobacco plantation workers to the demands of road construction. The people of the pueblos of Orizaba could be found simply by going to the work sites of the Camino Real.[76] This proposal would have mixed Nahua draftees with free-wage laborers of all calidades, and, indeed, a subsequent exchange suggests that some version of the proposal may have been implemented. Two years later a hacienda administrator complained to the subdelegate that his peónes had left to work on the road. After the subdelegate ordered the seven workers back to the hacienda, the road work director recommended that the day-laborer wage be raised to four reales to offset his labor shortfall. By approving the wage increase, the colonial state attempted to protect the tobacco haciendas' labor supply at a presumably low wage and still maintain a free-wage labor regime in road construction.[77]

The transition from indigenous village draft labor to free-wage labor was accompanied by a wage increase, and by 1791 the going rate had again increased, only to decline a few years later. Conflicting indications can frustrate attempts to classify the labor market as either a buyers' or a sellers' market. As can be seen in the wage reductions occurring in 1793–94, however, the long-term trends of change seem to have increasingly favored employers. In addition to the large forces of unskilled day laborers, these employers also relied heavily on skilled tradesmen to complete their construction projects. The records from Xalapa, the Puente de Escamela, and the Toluca road contain a wealth of evidence about these tradesmen, the analysis of which sheds yet additional light on the contours of the general labor market in late-colonial Mexico. In the next chapter, we turn our attentions toward aspects of the society against whose background everything up to this point took place.

CHAPTER FIVE

PATTERNS OF PEOPLE'S LIVES

SOCIETY AND FAMILY IN ORIZABA, 1777-1791

Whatever tranquility don Patricio Fernández may have enjoyed that day was suddenly shattered by the commotion at his door. As subdelegate of Orizaba, Fernández dealt with a host of issues and the full gamut of His Majesty's subjects, perhaps none as troublesome as the individual just then loudly demanding to see him right away. In the doorway loomed don Marcos González, a peninsula-born Spaniard with a militia captain's commission, a prominent tobacco planter then serving as an *alcalde ordinario* on the cabildo of Orizaba—with an attitude to match. Subdelegate Fernández may well have expected some sort of fracas. Residents of the Nahua barrio of Santa Gertrúdis had earlier complained that stone cut from within their barrio had been diverted for street repairs in Orizaba, in violation of the directive that authorized quarrying stone only for the construction of the Puente de Escamela. Fernández acknowledged the validity of the complaint and ordered that all cutting of stone cease immediately.

González vented his full fury at Fernández, accused him of deliberately obstructing progress on the all-important Puente de Escamela, and informed the subdelegate that he had personally countermanded the work stoppage order. Fernández retorted that as subdelegate he was the only official in Orizaba with any authority over the Nahua and then upbraided González for overstepping his bounds. Still unrepentant, González fulminated on and complained that he was sick and tired of the special privileges enjoyed by native peoples in New Spain and apparently also personally insulted Fernández. The subdelegate told González that whatever personal opinion the latter may have of him as an individual, as a royal official he would brook no disrespect whatsoever to his office, and then he dismissed González.

Knowing well that the matter was far from over, Fernández rode to Veracruz to consult with his immediate superior, the intendant of Veracruz. Meanwhile, González wrote a letter directly to Viceroy Revillagigedo, in which he laid out his complaints and frustrations. He got nowhere with the viceroy. Instead, he received a letter of reprimand, in which Revillagigedo advised him in no uncertain terms that insubordination to the subdelegate was a serious offense that would not be tolerated again. In the meantime, Subdelegate Fernández returned to Orizaba and resumed the duties of his office.[1]

JURIDICAL CONCEPTS AND THE CENSUSES OF 1777 AND 1791

This splendid anecdote reminds us of a crucial aspect of Spanish colonial society in general and of New Spain in particular: subjects of the realm belonged to either the república de yndios or the república de españoles, entities that coexisted in time and often in space. Members of the former juridical entity were ostensibly only of indigenous ancestry, whereas any African or European genealogy effectively placed one in the latter category. Yndios could live in towns or villages alongside nonyndios, as did hundreds of Nahua in Orizaba according to the 1791 census manuscript.[2] Or they could live in pueblos and barrios of their own, such as Santa Gertrúdis, in which only yndios had rights to land and resources.

Nahua pueblos could of course exist in geographical isolation, insulated in some ways from some of the intrusive aspects of the colonial regime. But depending on location, such villages could find themselves in daily contact and conflict with those same colonial influences. Located right in Orizaba, the Barrio de Santa Gertrúdis obviously fell into the latter category. As a center of tobacco production for the royal monopoly and a town connected with Mexico City and with Veracruz and Spain by what (sort of) passed for a highway, Orizaba was relatively unisolated, which affected the Nahua villages around it. The pueblo of Ixhuatlan, for instance, participated in the royal tobacco monopoly by renting land to it, and then the villagers cultivated the crop itself. In 1781, Ixhuatlan rented its land for four hundred pesos, but by accepting the contract, the pueblo gave up any and all rights to reclaim control of that land, for any purpose, for the duration of the contract. This 1781 contract authorized higher payments than the growers had received for the previous year's crop, the Crown apparently responding to the growers' declaration that they would refuse to plant unless guaranteed a reasonably profitable price.[3] The Ixhuatlan villagers would thus have benefited from this presumably lucrative contract, but at the cost of waiving legal protections applying to yndio land. Recall from chapter 2 that in that same year near Xalapa the Totonac villagers of San Miguel del Soldado failed in their litigation to prevent construction of the Camino Real through their maize field. The impact of colonialism on village society depended on time, place, and circumstance, as did the value of the special privileges of which don Marcos González so vociferously complained.

Although political economy is of crucial importance, there is still a great deal more to life. The artisans and laborers who built the roads and bridges of eighteenth-century Mexico obviously had more to their existence than their doings at the various work sites. Not only did they live as parts of a society, but each also played a number of roles in that society. Any number of categories can serve as bases for description and analysis of various aspects of their lives: position within a family structure, position within the official ethnic classification system (sistema de castas), membership in a militia unit, or position in a class structure, to name a few. Of course, finding documents that give voice to the eighteenth century's silent and illiterate masses is a difficult problem for the historian. In the case of our road-construction workers, census manuscripts contain much data in the categories named earlier and thus shed some light on how the masses might have lived their lives. This chapter is based on census records and provides a social background to our picture of laborers at work building the Puente de Escamela. Wherever available in the sources, the bridge workers themselves serve as the examples that support the particular argument being made. At some points in this study, persons not on the Puente de Escamela documents had to be chosen, but their cases still represent aspects of colonial Orizaba society that logically would have also applied to the typical bridge laborer. In an implicit sense, then, these latter cases are still connected to the bridge laborers even though a documentary link by name could not be established concretely.

Manuscript census enumerations prepared during 1777 and 1791 survive from both Orizaba and Xalapa.[4] These four documents have appeared in earlier chapters, and two of them serve here as the basis for analysis and description of Orizaba society itself. Unfortunately, I had to exclude the Xalapa census manuscripts from this part of the study for two reasons. First, the two Xalapa manuscripts were prepared seven and twenty-two years, respectively, after the latest construction payroll records from the Xalapa area that were studied in chapter 3, thereby making individual record linkages of nonelite persons virtually impossible and certainly unreliable. Second, the 1791 Xalapa census takers omitted the calidad (ethnic classification, such as "mestizo") of many individuals. The very omission itself is undoubtedly significant, but it renders impossible certain analyses that are facilitated by the apparently more meticulously prepared records from Orizaba, which nevertheless have weaknesses that must be kept in mind.

The census of New Spain ordered by Viceroy Revillagigedo identified individuals for purposes of conscription and taxation. The Orizaba volume carries an approval signature dated November 16, 1791, which, as noted earlier, falls within the period of construction of the Puente de Escamela, and it contains seventy-three nominal record linkages with the bridge payrolls. The Revillagigedo census contains no enumeration of yndios except for those married to nonyndios, although it does list those addresses in Orizaba at which none but yndios resided. The 1777 manuscripts list adult Nahua, but only the numerical totals given for Nahua children of the villa of Orizaba itself appear exact. The figures for the *cabecera* (head town) and barrios of the neighboring Curato de Santa María de Zoquitlán appear suspiciously round, such as the three hundred *párvulos de doctrina* (somewhat older children) and two hundred *párvulos de pecho* (nursing-age children) in the cabecera itself.[5]

Not only do we find whole segments of society left unrecorded along with questionable totals of others, but the totals may also reflect a substantial undercount of males older than the age of thirteen years. The 1791 Orizaba census lists 2,987 nonyndio males and 3,684 females living within the villa itself. The 1777 totals are 1,051 males and 1,183 females. The resultant male-to-female ratios are 0.807 and 0.888, respectively. These data, then, suggest a "shortage" or absence of men.

What happened to the 15 percent of the male population that would be needed to raise the 1791 Orizaba ratio to 0.95, 5 percent short of parity?[6] Perhaps they died, leaving behind the large number of enumerated widows, of whom more is discussed later in this chapter. However, there was no war to deplete the number of men, and Orizaba was not a port town that would have had many men away at sea. Diseases would have affected women also and brought down the numbers in roughly equal proportion. Perhaps a significant number of men migrated elsewhere in search of work, which is the conclusion that Cecilia Rabell Romero reaches in her analysis of 1777 Antequera. Martin Minchom suggests that a similar trend in Quito, in present-day Ecuador, resulted from the evasion of the tribute required of males and an influx of rural females for domestic service.[7] Because the 1791 census identifies a large number of artisans as natives of other places, Orizaba seems to have been at the time a destination rather than a source of migratory workers. Although there was no war just then, the threat of war between revolutionary France and Spain and its empire clearly existed by 1791.[8] Christon Archer perceptively suggests that men may have avoided enumeration in order to escape conscription and notes that the greatest inaccuracies most likely lay in the lower social strata.[9] Even with the undercount, the census data can still shed much light on colonial Mexican society. One can, of course, question the truthfulness of the entries, but as David A. Brading notes in his study of the contemporaneous census from Guanajuato, there would be little reason for most people to lie. Also, even though the census was the intendants and subdelegates' responsibility, local clergymen probably played a major role in its preparation. These secular priests would have known most of the people in a place the size of Orizaba and probably referred to their own parish records as they prepared the padrón manuscripts.[10]

The census taken in 1777 by the church for Viceroy Antonio María de Bucareli can be subjected to many of the same criticisms as the Revillagigedo census because Spain was then at war with Great Britain. As noted earlier, this 1777 document also shows a pronounced imbalance in the male-to-female ratio for the Orizaba area. But here again, these earlier documents can nevertheless tell us a great deal about society in late-eighteenth-century Mexico, particularly when used in conjunction with the 1791 data. Unless otherwise noted, the data appearing in this chapter are drawn from the 1791 Revillagigedo census from Orizaba and its environs and from the 1777 Bucareli census from the villa of Orizaba and neighboring Santa María de Zoquitlán.[11]

The original 1791 census manuscript reports a total of 9,119 nonyndios living in the Orizaba jurisdiction. An overcount appears evident as occasional arithmetic errors accumulated and were carried forward in the summary totals appearing at the top and bottom of each page. The tally from the database shows only 8,530 individuals identified in the census. The difference between the two totals may simply be the result of accumulated administrative errors, but another possibility tantalizes us at this point. The census takers may have intentionally increased the total by changing the numbers at places throughout the four-hundred-odd pages to raise the tally by about 600 persons. Recalling that our earlier concern that the imbalance in the male-to-female ratio stemmed from a "shortage" of approximately 600 males, we are now forced to consider the possibility that many men who were actually present on the streets of the villa were left off its census rolls, possibly having paid census takers to keep their names off lists that they knew would be used to conscript men for military service.

The seemingly large number of widows listed in Orizaba lends additional plausibility to the conjecture. The manuscript census lists 652 nonyndia widows, but only 133 nonyndio widowers. In perusing the manuscript, one is indeed struck by the frequency with which widows appear whose children are classed as *niños*, implying that the women themselves were comparatively young. Assuming for argument's sake that approximately 150 of the listed widows had indeed lost their husbands, approximately 500 women remain whose classification as *viudas* seems suspect. Returning these 500 ghost husbands to the male-to-female ratio raises it to 0.96, within 5 percent of parity. It is unclear what reasons the census takers would have to ensure the accuracy of the total number. Future research is required in order to examine more rigorously the preceding hypothesis and to determine if records from other locales exhibit similar tendencies.

When using the 1791 Orizaba census at the current state of knowledge, one must be careful to note that some numbers are understated, especially nonelite males between the ages of thirteen and forty. The situation has changed little since David Brading and Celia Wu argued that Mexican historical demography was in such a "rudimentary" state that simple presentation of data had its value.[12]

To best estimate the population in 1791 of the villa of Orizaba and its surrounding province, we can extrapolate Susan Deans-Smith's findings to a total of between 15,000 and 17,000 persons, of whom approximately 45 percent were classified as yndios.[13] Because the computer database contains a great majority of the nonyndio population, the analyses that follow here are best understood in a comparative context with the many fine historical studies of other locations that are also based on manuscripts from the Revillagigedo census.

CALIDAD AND SOCIETY

Whether colonial Mexican society can be characterized as being ordered by estate or by class has been the focus of a long-running historiographical debate. John K. Chance and William B. Taylor contend that estate, or caste barriers, had eroded to the point that late-eighteenth-century Mexican society could be better understood as being hierarchically arranged along lines of socioeconomic class. They derived their conclusions largely from the Bucareli and Revillagigedo census manuscripts from Antequera, present-day Oaxaca city. Thus, their sources parallel those used in this study of Orizaba. Robert McCaa, Stuart B. Schwartz, and Arturo Grubessich dispute Chance and Taylor's analytic methods and conclusions, however, contending instead that the society of Mexico remained a pigmentocracy, with one's place in the pecking order owed principally to skin color and the degree of European ancestry one could claim.[14]

The villa of Orizaba itself experienced a sizeable shift between 1777 and 1791 in the distribution of its population according to calidad, the ethnic classification of an individual within the official sistema de castas. The Revillagigedo census was prepared in one manuscript for the political jurisdiction of Orizaba that was established under the 1786 law creating intendancies in the Viceroyalty of New Spain. The 1777 census record books were submitted separately by each church parish in the Diocese of Puebla to the bishop, and for several geographic areas enumerated in the 1791 census the manuscripts from 1777 do not survive. This absence precluded comparing such outlying jurisdictions as the pueblos of San Miguel Tomatlán and Maltrata and the Hacienda de Tuxpango, with its labor force of 114 slaves, the only slave plantation left in the province in 1791.[15] In order to overcome the problems posed by the gaps in the records and to maintain as tight a basis for comparison as possible, I eliminated outlying barrios from this particular consideration and used only data from the villa of Orizaba itself to prepare table 5.1.

Table 5.1. Nonyndio Population of the Villa de Orizaba by Sex and Calidad

Calidad	1777				1791			
	Male	Female	Total	(%)	Male	Female	Total	(%)
Española[a]	205	196	401	(18.0)	1,630	1,818	3,448	(52.2)
Castizo	189	209	398	(17.7)	286	302	558	(8.8)
Mestizo	631	752	1,383	(62.0)	798	1,182	1,980	(30.0)
Pardo[b]	26	26	52	(2.3)	191	235	426	(6.4)
Unknown					64	135	199	(3.0)
TOTAL	1,051	1,183	2,234		2,969	3,672	6,641	

Sources: AGN Padrones, vol. 19, and AGI México 2580.
[a]Includes *españoles europeos* and *hijos de algo* for 1791.
[b]Includes *negros* and *morenos*.

The shift in percentages strongly suggests a general "whitening" of the population, but these data alone are insufficient to come to definitive conclusions. Orizaba's population grew between 1777 and 1791, and every number in the 1791 totals is greater than it had been in 1777. An additional complication arises in that in 1791, Spaniards born on the peninsula were denoted as *europeo* or *español europeo*. One hundred twenty-seven men and four women are listed as peninsula born, and another nine men and two women were classified as *hijo de algo* ("son of somebody") instead of as europeo or español europeo. The 1777 census manuscript, however, makes no distinction between creole and peninsular Spaniard. As is shown later in more detail, many of the peninsular Spaniards appear in both manuscripts. To facilitate quantitative comparisons and analyses, I created a single category by combining the 1791 totals of españoles, europeos, and hijos de algo. This is by no means intended to assert that no status difference existed between Spaniards of European birth and those of American birth.

The changes between 1777 and 1791 in the distribution of Orizaba's population by calidad appear to result in large part from two processes. The first is the application of the rules of the sistema de castas to the children of "mixed" marriages, and the second is what some historians have called "racial drift," when individuals appear in different documents with different calidades assigned to them. In the case of Orizaba in 1791, both processes are clearly evident in the census manuscripts.

The sistema de castas established a hierarchy of classifications based on a person's degree of European, African, and yndio ancestry. The system found itself in a sense undermined by the growing number of "hybrids." As codified in the eighteenth century, eighteen ethnoracial categories came to be prescribed, and the system was designed to preserve the high status of white skin and European ancestry.[16] Whether or not they intended to leave breaches in the administrative barriers separating the different calidades, the officials and priests who produced the 1791 Orizaba census paradoxically subverted some of the goals of the sistema de castas. By not using all of the prescribed categories, they left open paths of status improvement in a system intended to keep people subordinated in the ethnic hierarchy.

According to the official system of designations, the child born to a mestizo-yndia couple was to be labeled a *coyote*.[17] This designation can only have been intended to identify *castas* (persons of mixed ancestry), who could not claim that a full half of their progenitors' blood flowed from Europe and therefore could not be raised to the same status as a mestizo. But the census takers of Orizaba did not apply this rule, as shown by several cases from the manuscript.

An example of the actual practice is forty-one-year-old Antonio Alvino Flores, who lived in 1791 with his yndia wife at Calle de Nuestra Señora de Guadalupe, number 23. Because the Revillagigedo census was intended for taxation and conscription of members of the república de españoles, yndio spouses' names were not recorded, only their calidad. The couple had three children, including two *doncellas* (daughters of at least twelve or thirteen years of age) and a thirteen-year-old son named Vicente. All three of the children were labeled mestizos, the same as their father. Antonio Flores's occupation is listed as labrador, which usually means an independent peasant or farmer. If Flores had been a peasant with no plot of land on which he normally worked, he probably would have been listed as a jornalero, or day laborer. This should not be taken to mean that Flores owned any land. Along with the 137 labradores residing within the villa itself, he more than likely rented a plot from one of the absentee owners of large landholdings.[18]

Flores worked two stints as a day laborer on the Puente de Escamela. The first was for five days during the week of August 28, 1791, and he returned there for a six-day period ending March 11, 1792.[19] His appearance on the payrolls buttresses the hypothesis that many of the day laborers at the bridge were agricultural workers who occasionally signed on to earn a little hard cash.

This particular family also appears in the 1777 census, and Antonio Flores is the only laborer from the Puente de Escamela who can be *confidently* identified in both census records. The earlier document lists him only as Antonio Alvino, with an unnamed yndia spouse and two children: María Josefa, two years of age, and Vicente José, one year old. The other daughter who appears later in the 1791 records was apparently not yet born. In the 1777 census, the calidad of children is not recorded, so we are left to speculate as to whether María Josefa and Vicente José would have been labeled coyotes, according to the rules, or mestizos, as they were in 1791.[20] In either case, Antonio Flores and his family constitute a clear case of administrative "whitening."

Another example of this process can be seen in a family residing in Orizaba's Barrio Xalapilla. Justo Sánchez, one of the very few yndios whose actual name appears on the census manuscript, lived there with his mestiza wife, Estefania Tixtla. Their five children are also labeled as mestizos.[21] No person of any age is listed in the 1791 Orizaba census as a coyote, so equating persons having three yndio grandparents with those who had two appears to have been standard practice.

The castizo classification denoted a person with one español and one mestizo parent, or, obviously, one whose parents were both castizos. A number of historians have equated them with mestizos in order to include them for quantitative analyses in a category of persons who were not "all-white."[22] Such an approach serves some analyses very well, but it is not used in this study because it tends to mask the important role that the castizo category played in Orizaba's changing concepts of collective identity and self-identity. To borrow a phrase from Carl N. Degler's study of race in Brazil, a "castizo escape hatch" was operating in Orizaba.[23]

The rules of the sistema de castas clearly state that the child of español-castiza marriages are español. They also declare the offspring of a castizo-mestiza marriage to be *chamizo*.[24] As was the case with the coyote label, nobody in Orizaba was classified as chamizo in either the 1777 or the 1791 census records.

The actual practice becomes apparent in cases such as that of José Quesadas, a thirty-five-year-old mestizo who lived in the Barrio de Escamela with his castiza wife, Ana García, and their four children: an unnamed doncella, two niños, and a fifteen-year-old son, Andrés. All of the children were recorded as castizos.[25] José Quesadas worked as a day laborer on the Puente de Escamela for five consecutive weeks beginning on August 21, 1791, and earned a total of twelve pesos and four reals. Andrés joined his father at the bridge during the two weeks beginning on September 11, during which he earned two pesos and one and one-half reals.[26]

Some exceptions do appear in the records, which show that in a few cases this benefit of the doubt did not go to the individual in question. Agustín Gómez was a thirty-year-old castizo baker. In 1791, he lived with his mestiza wife, Josefa Hernández, and their daughter at Calle Real Vieja, number 20. The daughter was listed as mestiza. It is interesting to note that Gómez himself had been classified as mestizo in the 1777 census manuscript.[27] Only three other castizo-mestiza and mestizo-castiza families appear in which the children were labeled mestizos, and one of those was the family of Andrés Rodríguez, another castizo who had been listed as a mestizo in the earlier census.[28]

In contrast to the earlier-noted castizo-mestiza unions and their twelve children stand another forty such families whose total of ninety-three children were labeled castizos. They comprise a significant majority, which strongly suggests that the preponderant trend was to let such children move up a notch in the sistema de castas hierarchy. To summarize, it is now clear that in Orizaba, the official rules were contradicted in actual practice. From yndio and mestiza came a mestizo, from castizo and mestiza came a castizo, and the rules already held that from castizo and española came an español. Given these practices, one can easily construct a family tree for a creole who was actually of half-yndio ancestry. The sistema de castas is at bottom a social construction, and the Orizaba census records are examples of the use of its hierarchy as a means of collective social self-definition. If a person claimed to be a creole, conformed to the norms of creole society, and was accepted as a creole by society, then that person was a creole, no matter how many yndio ancestors he or she may have had. In this manner, the operation of the castizo escape hatch played a key role in the "whitening" process.

The system operated much differently for Orizabeños of African ancestry. The administrative escape hatch described earlier did not work for them. As Patrick J. Carroll points out in his study of Afro-Mexicans in Veracruz, Xalapa, and Orizaba, one black ancestor usually meant that a person was designated *pardo*, which meant having a "grayish-brown skin."[29] Here again, we find substantial differer_es between the official rules of the sistema de castas and government and ecclesiastical officials' actions.

The only classifications appearing in the "Pardo y Moreno" section of the 1791 Orizaba census are pardo, *moreno* (ebony-colored), and *negro* (black). The first two do not appear at all among the official classifications, which describe a variety of possible mixtures, such as *mulato* (half white, half black), *zambo* (three-fourths black, one-fourth white), *morisco* (three-fourths white, one-fourth black), or *lobo* (half yndio, half black), to name a few. Orizaba may be unusual in this regard because a perusal of the Xalapa census reveals frequent use of the morisco and mulato classifications.[30] John K. Chance observed a similar pattern in the Antequera census, but also found that later parish records of marriages used the fuller array of categories.[31] The Orizaba census of 1791 contains 19 negros, 5 morenos, and 742 pardos, which suggests that most persons of African descent also had many yndio ancestors. Gonzalo Aguirre Beltrán notes that the term *mulatos pardos* commonly referred to such people and that they constituted by far the most numerous group of African castas in colonial Mexico.[32]

The family of Miguel Pizarro very clearly illustrates the process. A sixty-year-old pardo blacksmith, Pizarro lived at Calle Tercera Real, number 74, with his wife, Ana Chacón, who was also a parda. In 1777, they had two daughters living with them. In 1791, the daughters lived nearby, each with her respective husband. María Antonia Pizarro had married Manuel Rendón, a creole blacksmith who was thirty years old in 1791. They lived at Calle Tercera Real, number 77, with their young son and daughter. María Antonia's sister, Ana, had married an yndio and lived with him and their young daughter at Calle Tercera Real, number 79. Of course, Ana's husband's name and occupation do not appear in the census record, but one plausibly surmises that he, too, was a blacksmith. The census taker classified each of the three young cousins as pardo.[33] The important question of marriage preferences is considered in detail later, but suffice it here to say that one avenue of movement "upward" through the racial hierarchy was kept closed to the pardo population of Orizaba. Incremental movement by succeeding generations was, however, by no means the only path to individual status improvement within the sistema de castas. As with yndios, there may have been valid reasons for some pardos not to attempt to pass.

Ben Vinson III has analyzed the roles that the militia played in the pardo community, especially in the regions around the Gulf of Mexico coast. He finds that the militia became a source of racial identity for pardos, one that carried substantial prestige and tangible benefits such as the military *fuero,* whereby a militia person escaped jurisdiction of civilian courts for any alleged offense. Vinson shows that militia service offered a path for social advancement through an officer's commission and acquisition of other status indicators not usually available to pardos who were not in the militia.[34] The Orizaba census confirms many of Vinson's findings for other places. The record shows 154 militiamen, with 22 of them listed as pardos, or 14.3 percent of the total. This is more than twice the proportion of pardos in the overall population. Fourteen of the pardos were cavalry troopers of the Veracruz Lancers, including one corporal. Eleven Orizaba men of other calidades also served in this regiment. The remaining militiamen belonged to a local infantry unit that would later become the Regiment of the Three Villas. This total included six lieutenants and four captains, all españoles, four of whom were peninsula born. That no pardos of Orizaba held royal commissions should not be interpreted to mean that militia service offered none of the opportunities that Vinson found in Puebla and elsewhere. Exemption from tribute and the military fuero by themselves doubtless counted for a great deal.

SOCIAL CLIMBERS AND CHANGES IN CALIDAD

Sometimes it was possible for persons to get their own calidad changed on official records. R. Douglas Cope finds such practices in central Mexico City's Sagrario Metropolitano parish in the late seventeenth century, which may have presaged trends elsewhere in New Spain over the course of the eighteenth century.[35] Chance and Taylor contend, for instance, that by 1793, many mulatos were successfully passing as creoles in Antequera, the present-day city of Oaxaca.[36] Census manuscripts suggest that the same situation existed in Orizaba then.

I tested this hypothesis using the computerized database of the 1777 and 1791 Orizaba censuses mentioned earlier. I queried and manipulated the database in a variety of ways in order to identify individual persons who appear in both documents. If the names were the same, then the record became a candidate for nominal linkage. If more than one person in the other census manuscript shared the name under consideration, the ambiguity could sometimes be resolved by comparing other data, such as the spouse's name, or the person's age if it was recorded in both manuscripts. Ambiguities often still persisted, which mandated caution. Logic clearly dictates that if, for example, three women named María Gómez appeared in each of the two censuses, then in all likelihood they were not six different persons. In such cases, however, prudence demanded that the records not be correlated or linked in the database.

The results of this study thus understate the number of persons whose names were actually recorded in both censuses. An effort to identify yndios from the 1777 census who were passing as castas (of mixed ancestry) in 1791 had to be abandoned after it yielded inconclusive results. Two historians note a decline in the numbers of yndio families in Tepeaca between the censuses of 1743, 1777, and 1792, but they do not raise the issue of "passing." Another finds that yndios frequently claimed to be mestizos in Mexico City in order to avoid tribute. A fourth scholar argues that yndio migration to town often resulted in a change from yndio to casta status, and if one accepts the notion that any higher-ranking calidad had some value to common people, it seems logical that such cases would also have occurred in Orizaba. Such cases would appear in the census records, but we must remember that the threat of military conscription might have made many men more likely to report themselves as yndios to the officials and priests who prepared this particular census enumeration.[37]

In total, 1,012 persons were linked between the two census reports. Of those, 330 persons were listed in 1791 as having a calidad that differed from that listed in 1777. The vast majority of the persons whose calidad changed from the earlier listing moved "up" the casta hierarchy. There, 260 "social climbers" outnumbered by almost four to one the 70 people whose status declined. Table 5.2 displays the number of these changes in calidad between the two documents.

There appears to have been a fairly even split with respect to gender among those moving up and those drifting downward, with two exceptions. In the first, pardo men and women changed in roughly equal numbers, but no pardas became españolas. Three pardo men became creoles, but it is impossible to draw significant conclusions based on only three of anything. Two of them were a father-and-son pair. Interestingly enough, the father and the third man were both married to castiza women in the 1777 records,

Table 5.2. Changes in Census Calidad by Gender in Orizaba Between 1777 and 1791 among Persons Identified in Both Census Manuscripts

	1791 Calidad			
	Español	Castizo	Mestizo	Pardo
1777 Calidad				
Español	—	12 (3,9)	14 (3,11)	0
Castizo	93 (44,49)	—	33 (15,18)	7 (4,3)
Mestizo	133 (60,73)	21 (12,9)	—	4 (2,2)
Pardo	3 (3,0)	2 (1,1)	8 (3,5)	—

Sources: AGN Padrones, vol. 19; AGI México 2580.
Note: Figures enclosed in parentheses indicate the number of males and females, in that order, composing the total immediately above them.

but all four members of the two pardo-castiza couples show up in 1791 as creoles. One of the couples, Miguel Valderrama and Antonia Bustamante, had a daughter in 1777, but she could not be identified in the later census. The other couple, barber Cristóbal Campos and Nicolasa Vega, lived at Camino del Calvario, number 11, with their sixteen-year-old son, Ygnacio, who was employed as a purero (cigar maker) at Orizaba's Royal Tobacco Factory.[38] None of the three men's occupations seems so prestigious that it alone would have caused the rise in calidad, so the underlying reasons remain at this point obscure.

The other exception is that of downward-drifting española women, of whom there were twenty as compared with only six men. Two of the women and one of those men were married to mestizos in 1791, but these are the only three of the total of twenty-six creoles whose downward drift can be attributed to a match-up with the lower calidad of a spouse. Additionally, five castizo men and four castiza women were downgraded to match their mestizo spouses. Here, the trend in Orizaba runs counter to Robert McCaa's findings for Parral, a silver-mining town in northern Mexico. McCaa linked parish records with a census manuscript and found that women's status depended on that of men, changes in calidad usually occurred to match that of a spouse, and "[m]ost racial drifting was downward."[39] For Orizaba, however, the two census manuscripts show a marked tendency for upward movement of persons paired with mates of the same calidad in 1791.

Of the 260 persons whose calidad improved, 50 were in endogamous marriages with a partner who was not identified as being of a different calidad in 1777. One parda woman became a mestiza, and one mestiza became a castiza. Twenty-five mestizos became creoles, including eleven men and fourteen women. Nine castizo men and fourteen castiza women also became creoles. It should be noted that the numbers of both upward and downward movers include some couples who both changed calidad but could not be identified because one of the two partners could not be identified in both census manuscripts. Thus, the true significance of the tallies lies in how they compare to each other and not in the actual numbers, and they emphatically suggest that brides and grooms were more likely to move "up" rather than drift "down" to calidad equality with their spouses.

Seventy-five married couples appear in the two Orizaba census records in which one partner's calidad changed. Only fifteen of those changes resulted in that couple's appearance in the 1791 manuscript as an endogamous pairing. Five of these pairings resulted in mestizo couples and resulted from downward movement on the parts of two castizo men, one castiza woman, and two creole women. The other ten involved upward movement. One parda woman became a mestiza. Seven mestizos (five men and two women) and two castizo men became creoles and thus matched the creole status of their nine respective spouses. These numbers emphatically suggest that both brides and grooms were more likely to move "up" than "down" to calidad equality with their spouses.

The impetus for the majority of the calidad changes remains obscure. Chance points out that some persons purchased certificates of *limpieza de sangre* (purity of blood) that may have been suspect, and he also shows that the census takers in Antequera simply accepted most people's self-declarations concerning their casta status, which indeed may have been the case in Orizaba as well.[40] Ann Twinam argues otherwise, noting that only thirteen pardos were "whitened" after 1795, based on petitions located in the Archivo General de Indias in Seville. Her study covers all of Spanish America, and she notes the paucity of petitions originating in what is today Mexico. She concludes that discrimination in Mexico must have been "extreme." One must naturally ask: How extreme as relative to what? Discrimination was certainly the very essence of the sistema de castas. But the small number of petitions from Mexico may also mean that they were unnecessary in many cases of casta promotion, so perhaps the degree of discrimination may have actually been less severe than in places such as Cuba and elsewhere in Spanish America.[41]

Twinam's important study of petitions for *gracias al sacar* (the official name of the actual document) suggests that such certificates of status were few and far between. She located 244 petitions supported by many informative documents such as legal pleadings, affidavits, and so on. All of these documents shed a great deal of light on the petitioners' lives and allowed Twinam to thicken her descriptions of them substantially to a degree not permitted by the thinner data of the census manuscripts. This pinpoints a difficult problem in studying the social history of nonelite persons, which is that these people were very unlikely to leave archival paper trails as rich as those marking the life courses of Twinam's petitioners. Many of her subjects were trying to paper over questions about their ancestries in order to establish eligibility to hold high-ranking posts such as *regidor* in a particular cabildo. As demonstrated later in this chapter, most of the social climbers in Orizaba had occupations for which it is inconceivable that they would have been able to pay the five hundred to eight hundred reals required to file a petition for a certificate of limpieza de sangre. If Twinam's prodigious research efforts uncovered all of the extant petitions, then more people "purified" their census calidad between 1777 and 1791 in this one secondary, provincial Mexican city than petitioned for gracias al sacar throughout the Spanish American Empire. If we invoke an idea proposed by Charles Tilly, the analytic significance of the Orizabeños' successful claims to higher calidad lies in the sheer number of cases, which tends to offset the comparatively threadbare data contained in a census manuscript.[42] But for nonelite persons, those threadbare data are usually all that we are lucky enough to get.

R. Douglas Cope contends that personal ties counted for more than calidad in important daily matters such as access to credit and so on. He also finds that in late-seventeenth-century Mexico City, creole status imputed benefits to an individual, but there existed little practical difference between the other calidad categories.[43] How important was calidad status to nonelite persons? After all, we have seen that one of Miguel Pizarro's parda daughters married a creole, and the other married an yndio. They all lived in close proximity on the same street. Differences in calidad surely cannot have meant much on a daily basis in their lives. One might wonder how someone such as Cristóbal Campos benefitted from his social promotion. After all, in 1791 he was only a barber and still married to the castiza Nicolasa Vega. Status must have been for Campos a largely intangible thing, but it seems a mistake to presume that it counted for little among those who had little of it.

Creoles still accounted for only approximately one-third of Orizaba's total populace in 1791. Even so, calidad, or social constructions of "race," must have remained a crucial determinant of Orizabeños' perceptions of their individual and collective selves. The mere facts that the census takers so carefully recorded calidad and that the viceregal census instructions so mandated testify to the extreme importance of ethnic hierarchy to eighteenth-century Mexican society.

Marriage Preference and Family Structure

The configuration of colonial Mexican society owed much to the family unit in general and to marriage choices in particular. The positions of both sides in the previously discussed historiographical debate over the relative importance of class and calidad rest in large part on analyses of marriage choices and focus on endogamy or exogamy with respect to partners' calidad. The distinction between endogamy and exogamy of course depends on how one draws the distinctions between groups.

The sistema de castas as utilized by the census takers themselves serves as the basic referential framework for what follows here. Some modifications to the enumerated calidades were appropriate. Because the total of 19 negros and 5 morenos was so small in comparison with the 742 pardos in the 1791 Orizaba census, the three groups were combined into one category for the purposes of this study. The 1777 census manuscript contains no distinction as to the continent of birth of españoles, but in 1791, 142 men and 5 women were identified as of European birth. Interestingly, don Pedro Barroso is listed as having been born in the Canary Islands, and he is identified only as español, not as español europeo. Also, a handful of prominent persons, such as don Patricio Fernández, subdelegate of Orizaba, are listed as neither creole nor peninsular Spaniard, but only as hijo de algo.[44] In classifying a marriage as endogamous, American-born creoles, peninsula-born Spaniards, and hijos de algo were combined into a single category of españoles. This establishment of exactly parallel categories facilitates direct comparison with data extracted from the 1777 census. Because there were so few European-born women in Orizaba, combining the three categories also avoids the automatic classification of those marriages between Spanish-born men and creole brides as exogamous and implicitly less than desirable for the partner of lower standing.

Calidad equality between husbands and wives characterized the majority of Orizaba's families in both censuses. Table 5.3 demonstrates the strength of that tendency for all of the couples in the 1777 census. Table 5.4 does the same for the 1791 manuscript, of course without a figure for yndio couples. Given the upward mobility within the sistema de castas seen previously, the greatly increased number of creole couples is not surprising.

Table 5.3. Calidad Distribution of Husbands and Wives in Orizaba, 1777

| | Wives | | | | | |
	Española	Castiza	Mestiza	Parda	Yndia	Totals
Husbands						
Español	72	4	42	7	0	125
Castizo	18	55	82	2	0	157
Mestizo	36	47	568	8	15	674
Pardo	2	16	4	14	5	41
Yndio	3	0	27	7	1,362	1,399
Totals	131	122	723	38	1,382	N = 2,396

Source: AGI México 2580.

Table 5.4. Calidad Distribution of Husbands and Wives in Orizaba, 1791

| | Wives | | | | | |
	Española	Castiza	Mestiza	Parda	Yndia	Totals
Husbands						
Español[a]	589	30	112	14	3	738
Castizo	25	33	37	1	1	97
Mestizo	71	15	343	22	31	482
Pardo	18	3	46	56	10	133
Yndio	7	1	18	4	—[b]	30
Totals	710	82	556	97	42	N = 1,490

Source: AGN Padrones, vol. 19.
[a]Includes españoles, españoles europeos, and hijos de algo.
[b]Total not available from the 1791 census owing to nonenumeration of yndios.

As mentioned earlier, calidad endogamy preference is a key point of contention in the estate versus class debate of colonial Latin American historiography. At the center of that debate lies a dispute over which methods of statistical analysis yields more accurate interpretations of information, such as the endogamy data presented in tables 5.3 and 5.4. The contribution of this study to that particular dispute is unfortunately reduced by the lack of any record of the number of yndio marriages to complete the matrix. One can reasonably surmise that the number might be between one thousand and fifteen hundred, given the estimated percentage of yndios in Orizaba's total population. Caution protests, however, against basing mathematical analyses on what can only be an arbitrary guess, no matter how educated it might be. The gap in this record regrettably precludes a full statistical comparison of trends in endogamy for Orizaba.

Many studies of colonial Latin American society have used families as the subjects of analysis.[45] Both of the Orizaba census records are organized into what appear to be family units. The 1777 manuscript has no street address data, a situation that requires the historian to accept the households as defined by the census taker, which probably masks a significant number of families who in this study's analytic scheme would be counted as "extended." In parts of the 1791 census, one can see separately identified family units with surnames in common living at the same address. Such a situation implies family extension instead of the nuclear families that census takers often recorded.

That part of the 1791 census covering the Villa de Orizaba itself contains street addresses and house numbers, with the exception of the Barrio de Santa Anna. My analysis of family structure is thus based on only that part of Orizaba for which it can be determined exactly who lived at what address. This approach allows a more realistic assessment of the propensity for living in extended families.

To illustrate the point, consider the situation at Calle de Nuestra Señora de Guadalupe, number 48. The padrón identifies it as the residence of three nuclear families, each consisting of a husband, a wife, and some children. A closer look, however, raises questions about these three supposedly separate families. A fifty-three-year-old creole blacksmith, Eusebio Antonio Saquero, lived there with his española wife, Josefa Sánchez, nineteen-year-old son Francisco, thirteen-year-old son José, one younger son, and three younger daughters. Also in residence is another creole blacksmith, twenty-eight-year-old Miguel Modesto Saquero; his mestiza wife, María Guadalupe; and their three young daughters. The picture is rounded out by the third married couple, twenty-three-year-old José Antonio Saquero and Anna García, who also lived at the same address with their young son. Including the two teenage sons, each of the five men was a blacksmith, and it seems highly unlikely that they were not related to each other.[46] In construction of the computerized census database, I combined families such as the Saqueros and counted them as one extended family if they lived at the same address. Because the Saquero family consists of parents, children, children-in-law, and grandchildren, it is considered to be vertically extended. The presence of another blood relative outside the direct generational chain would make it a horizontally extended family. A nuclear family consists of a married couple and their minor children.[47] For this study, twenty-five was taken as the age of majority. Because unmarried, mature daughters appear only as doncellas with no age given, the number of extended families in Orizaba is to a degree understated in this study.

Most Orizabeños lived in nuclear families, although quite a few were single and did not live with any kin. Approximately one-fourth of the residents of the villa lived in extended family units, as can be seen in table 5.5, whose numbers include combinations of nuclear families as described earlier.

The 1777 census manuscript only rarely lists families in an adjacent numerical sequence with the same surname. The families are listed one by one, arranged by street but with no house numbers. So seldom do

Table 5.5. Distribution of Family Structures by Type in Orizaba, 1791

Type Family[a]	Number of Families (%)	Number of Persons (%)
Solitary	865 (28.8)	865 (14.8)
Nuclear	1,628 (54.2)	3,625 (60.9)
Vertically Extended	173 (5.8)	392 (6.6)
Horizontally Extended	335 (11.2)	1,054 (17.7)

Source: AGN Padrones, vol. 19.
Note: Barrio de Santa Anna excluded from study because house numbers not in census manuscript.
[a] Families at one address sharing a common surname counted as one family.

Table 5.6. Distribution of Family Types and Individual Residences in Orizaba, 1777 and 1791

	1777	1791
Percentage of nuclear family units	82.2	76.2
Percentage of extended family units	17.8	23.8
Percentage of persons in nuclear families	78.9	71.5
Percentage of persons in extended families	21.1	28.5

Sources: AGN Padrones, vol. 19; AGI México 2580.
Note: Solitary persons not included.

sequential entries for family members have common surnames that it is likely that the combinations entered into the computer database actually place the 1791 totals on a comparable basis with the 1777 counts. Between 1777 and 1791, the percentage of extended families rose from 17.8 to 23.8 percent, and as can be seen in table 5.6, this change represents an increase from 21.1 to 28.5 percent of the total number of persons belonging to extended families. The higher numbers result from the tendency of extended families to have more members than did nuclear ones.

The tally of extended families by residence fits a standard convention, but it may understate the propensity for family extension because it possibly counts as nuclear families a number of networks that probably operated in the manner of extended families. Consider the Saquero families mentioned earlier. All the men worked as blacksmiths, and the family units lived in close proximity on the same street. It is highly unlikely that these groups existed in substantial isolation from each other. An illustration of Orizaba's Camino Real in the early nineteenth century shows the road adjoined by two rows of mostly single-story buildings with doorway after doorway, undoubtedly the residences of families such as the Saqueros. The picture suggests that the domiciles were small and that architecture may also have constrained family residential choices.[48] Although we will not depart from convention and start combining neighboring families with apparent connections for purposes of this study, we should keep such reservations in mind when analyzing family structure data from any location.

Approximately one-ninth (12.3 percent) of the seventy-three bridge workers who were linked to the 1791 census lived in extended families, of which two men's families were vertically extended. Considerably lower than the average, this statistic suggests that persons on lower rungs of the socioeconomic ladder were much less likely to live in extended families than were higher-status families.

José Vargas was a thirty-nine-year-old creole whose occupation is listed as operario, most likely a cigarette maker at the Royal Tobacco Factory.[49] He also worked on the Puente de Escamela during the week of August 28, 1791, throughout the month of January 1792, followed by one-week stints in each of the two following months.[50] Vargas lived at Calle del Encierro, number 9, along with his widowed sister, Ana María Vargas de Pérez, and her two sons. The elder son, José María Pérez Vargas, was a sixteen-year-old cobbler. Two mestiza widows and their families also lived at that address: María Lechuga, with two young daughters, and Matiana García, who had two minor sons. These apparently tight living arrangements suggest equally tight economic straits for the occupants, and one may reasonably speculate that José and Ana María's decision to share a residence was based, at least in part, on the hope for mutual economic benefit, if not outright necessity.

Additional research of census records is needed to place Orizaba in a more solid referential context. Questions such as whether these findings typify eighteenth-century Mexico might then be answered. Comparisons with other places in Europe and North America are enlightening. One might suppose that the families in a Mexican city would exhibit a greater tendency toward extension, and here, in fact, the numbers do support the generalization. It was the situation in the city of Guadalajara in 1821, which reported a higher incidence of extended and multiple families than in Europe and North America north of Mexico.[51] In both England and the United States, extended families accounted for somewhat less than 10 percent of family units at the end of the eighteenth century.[52] In the second half of that century, only 4 percent of families in the Castilian province of Cuenca were extended.[53] A similar analysis of fourteen villages near Santa María del Monte in the Spanish province of León found that in 1752 slightly less than 6 percent of families were extended.[54]

Historians such as Peter Laslett and Lawrence Stone have held that family extension became increasingly common in Europe in response to increasing economic pressures resulting from industrialization. In contrast, Steven Ruggles's data show that the propensity in Europe and the United

States for vertical family extension had always been high. Ruggles attributes the nineteenth-century rise in the incidence of extended families to decreased mortality (i.e., a longer life expectancy) and younger age at first marriage, which meant that more grandparents were still alive and available to form extended families.[55] In eighteenth-century Orizaba, men older than sixty also showed a heightened tendency to live in extended families, but a majority still lived in nuclear families. As can be seen in table 5.7, 33.3 percent of men older than sixty throughout the province lived in extended families, as compared with 24.3 percent of the population as a whole. (The latter figure is lower than the results presented earlier because it was extracted from only the Villa de Orizaba itself.) Unfortunately, older women could not be included in the analysis because the census takers did not record their ages. Even counting only older men, however, yields a much higher propensity for vertical family extension in Orizaba than was found in the studies of other locations, including Spain, the United States, and England.

Another pattern of clear distinction between family structures in Mexico and in Europe or the rest of North America emerges from an examination of the stem hypothesis. The stem family is a special case of vertical family extension in which one son, usually married, remains at home to take over the family farm or business. All other siblings leave the parents' household upon marriage and go off to form other households. The stem family was a means of conserving property in preindustrial agricul-

Table 5.7. Residence Pattern by Family Structure of Males Older Than Sixty Years, Orizaba 1791

		Solitary	Nuclear Family	Vertically Extended Family	Horizontally Extended Family
Men >60 yrs.		17	53	19	15
	(%)	(16.7)	(52.0)	(18.6)	(14.7)
Population as a whole	(%)	(14.8)	(60.9)	(6.6)	(17.7)

Source: AGN Padrones, vol. 19.
Note: Enumerated families at same address and sharing a surname considered to be one family.

tural societies. Ruggles concludes that the stem hypothesis itself is valid and finds a marked preference for the stem in eighteenth-century England.[56] William A. Douglass finds that the Basques of Navarra also preferred the stem arrangement, although its actual frequency was statistically limited by demographic constraints.[57]

In Orizaba, none of the vertically extended families could be clearly identified as of the stem type. The great majority of extended families contained more than one coresident sibling. In others, the adult son was engaged in a different occupation than the father. These latter cases numbered only eleven in all. Nor was there a single case of a son-in-law and daughter remaining at home without other siblings present. The Basque peasant stem family played an important role in conserving property, but the use of other strategies to deal with laws requiring partible inheritances in colonial Spanish America has been well documented. Particularly in elite families, a large dowry for a daughter whose husband was being brought into the family business was a common practice, but became somewhat less common as the eighteenth century wore on.[58]

Women played a variety of important roles in the family, including as heads of household. Approximately one-fourth of Guadalajara households were headed by women in 1821.[59] In 1777, Orizaba women were equally likely to occupy that principal role, assuming that the census takers' decision regarding whom to list first in the manuscript identifies the head of the household. In that document, 470 household heads were women, which is 29.3 percent of the total number of 1,607 household heads, which results are consistent with findings of 30 percent for Mexico City and 40 percent for Antequera in 1777.[60] By 1791, however, the number of female-headed households had declined to 346, or 18.3 percent of the total, for this later census. Neither of the Orizaba figures includes solitary women.

Questions such as changes in the incidence of female-headed households or preferences for certain family structures demand additional research as surviving documentation permits. Orizaba had a large proportion of female-headed households and a high incidence of family extension, but questions remain. Was the decline in the fraction of female-headed households in Orizaba indicative of Mexico in general? How many of the young widows in the census were actually widows, how many had husbands hiding from the census taker, and how many reported themselves to be widows but were instead unmarried mothers? One might generally presume that any Latin society contained more of both vertically and horizontally extended families than would be found in North America, but here again we have to ask: Is our finding for Orizaba unique or representative of Mexican society in general? As future researchers, we must count and not merely generalize if we are to take advantage of all opportunities to expand our knowledge of colonial Mexican society.

Social Class and Occupational Structure

The status barriers of the sistema de castas by now were apparently so permeable that Max Weber's concept of a closed estate clearly does not describe the system as it was actually applied. The tandem-operating "whitening" mechanisms suggest that, much as Chance and Taylor contend for Antequera, Orizaba was indeed becoming increasingly ordered by socioeconomic class because by 1791 more than half of the nonyndio population had español status. Consideration of class—or, indeed, of *which* particular concept of class—introduces categories of social differentiation that the Orizabeños of 1791 would not have used to describe themselves. It may seem problematic or even teleological to attempt to determine the relative importance of race or class differentiation because we must use analytic concepts created long after the existence of the social constructions of race or calidad used in the Spanish American world of the time. Noting but then setting aside for the moment any fears about denial of historical agency to the Orizabeños themselves, we turn to the issue of class.

In questions of class, an individual's occupation serves as a primary indicator. Prestige defies measurement and perhaps increases the difficulty of avoiding anachronism, but we must also take account of it and of wealth in order to understand the social position of persons such as royal officials or some clergymen. The 1791 census lists the occupations of most nonyndio men age thirteen and older, but lists no women's occupations at all. The presence of women in the labor force is undeniable, and some measure of it can be gained through considering occupations for which no men are listed at all, such as *cocinero* (cook) in the case of the Orizaba census. Many bakers and a few candy makers appear in the census, and because well-to-do Orizaba families undoubtedly employed cooks, one may safely assume that all of them were women or yndio males. The Royal Tobacco Factory was a major source of employment, and one historian finds that approximately half of its cigar and cigarette makers were women. Another has shown that many women worked in the factories in a variety of capacities.[61] Limitations of the source documents notwithstanding, analysis of that part of the occupational structure that is recorded still offers insights into Orizaba society.

In his study of 1821 Guadalajara, Anderson proposes several broad occupational categories and arranges them in a hierarchy. Distribution of the reported occupations of the 1791 Orizaba census versus calidad according to Anderson's scheme appears in tables 5.8 and 5.9. To illustrate the problematic nature of this hierarchy, one might consider that by counting Orizaba's jailer in the administrative elite, we have placed him higher on the social scale than several well-to-do tobacco farmers (labradores). Just as Chance finds for Antequera, the term *labrador* in Orizaba covered a wide range of wealth and economic power.[62] Perhaps the merchants and planters might better be considered along with the administrative elite as a single group in contrast with artisans, laborers, and servants.

Table 5.8. Occupational Structure of Orizaba Males, 1791

	Calidad				
Occupation	E	C	M	P	U
Group 1 (Administrative Elite)					
Regular clergy	23	0	0	0	0
Secular clergy	37	0	0	0	0
Other religious	5	0	0	0	0
Sacristan	2	0	0	0	0
Government executive	23	0	0	0	0
Government clerk	6	0	0	0	0
Customs guard	14	0	1	0	0
Postal official	4	0	0	0	0
Military professional	4	0	0	0	0
Attorney	4	0	0	0	0
Teacher	4	0	0	0	0
Jailer	0	1	0	0	0
Constable	1	0	0	1	0
Tobacco factory executive	11	0	0	0	0

Table 5.8. (continued)

Occupation	Calidad				
	E	C	M	P	U
Tobacco factory clerk	3	0	0	0	0
Tobacco factory guard	42	0	0	0	0
Messenger (*corredor*)	4	1	1	0	0
Night watchman	0	1	2	0	0
Granary keeper	1	0	0	0	0
Physician	3	0	0	0	0
Apothecary	2	0	0	0	0
Surgeon	1	0	0	0	0
Torero	1	0	1	0	0
Scribe	16	0	0	0	0
Student	39	0	0	0	0
Group 2 (Merchants and Planters)					
Sugar plantation overseer	4	0	0	0	0
Sugar plantation worker	4	1	7	10	0
Tobacco grower	38	0	1	0	0
Tobacco rancho overseer	10	0	3	0	0
Hacienda owner (unspecified)	10	0	6	0	0
Campista	3	0	2	0	0
Itinerant peddler	6	0	1	0	0
Merchant (*comerciante*)	28	0	0	0	0
Cashier (*sirviente*)	29	0	4	0	0
Petty trader (*tratante*)	125	5	28	2	0
Haberdasher (*tendero*)	5	0	3	0	0
Tobacco seller	4	1	2	0	0
Group 3 (Artisans)					
Master tobacco worker	2	0	0	0	0
Tobacco factory worker	7	0	3	0	0
Cigar maker/cigarette maker	103	4	15	4	0
Tobacco sifter	0	2	1	0	0
Tanner	6	2	2	1	0

Table 5.8. (continued)

Occupation	Calidad				
	E	C	M	P	U
Baker	6	7	13	3	0
Weaver	31	6	15	0	0
Cobbler	19	14	37	9	1
Hatter	1	0	5	0	0
Silversmith	20	0	1	0	0
Tinsmith	1	0	0	0	0
Barber	22	1	2	0	0
Confectioner	0	1	3	0	0
Fireworks maker	2	2	6	0	0
Saddlemaker	15	3	4	1	0
Leatherworker (*gamuzero*)	7	3	9	0	0
Beltmaker (*talabartero*)	1	0	2	0	0
Hairdresser	1	0	2	0	0
Button maker	3	0	0	0	0
Embroiderer	1	0	0	0	0
Delftware maker (*losero*)	9	0	4	0	0
Rosary maker	1	0	1	0	0
Gilder	1	0	2	0	0
Soap maker	2	0	0	0	0
Gunsmith	1	0	0	0	0
Slaughterman	2	0	0	1	0
Butcher	1	0	0	0	0
Pork butcher	2	0	2	0	0
Flagon maker	0	0	0	1	0
Tile maker	0	1	5	0	0
Carpenter	49	4	16	5	1
Blacksmith	47	4	16	5	1
Tailor	56	12	24	12	0
Painter	4	0	0	0	0
Sculptor	1	0	0	0	0

Table 5.8. (continued)

Occupation	Calidad				
	E	C	M	P	U
Musician	3	1	1	0	0
Candle maker	5	0	2	0	0
Dyer	1	0	0	0	0
Mason	2	1	3	1	0
Miller	0	1	0	0	0
Group 4 (Laborers)					
Sugar plantation worker	4	1	7	10	0
Tobacco ranch worker	26	11	112	9	0
Cattle herdsman	5	3	9	7	0
Ice man	0	0	1	0	0
Water carrier	1	0	0	0	0
Porter/deliveryman	9	0	6	0	0
Ox driver	1	0	0	0	0
Muleteer	29	4	23	4	0
Coachman/Litter driver	0	0	2	2	0
Day laborer (jornalero)	49	0	55	2	0
Farmer (labrador)	145	55	131	9	0
Tobacco factory operario	61	20	126	59	0
Group 5 (Servants and Slaves)					
Male servants (free)	18	4	34	9	1
Female servants (free)	40	4	142	45	4
Slaves	0	0	0	79	0

Source: AGN Padrones, vol. 19.
Note: Calidad codes: E = español, español europeo, hijo de algo; C = castizo; M = mestizo; P = pardo; U = unknown.

Creoles are distributed throughout the various categories, but peninsular Spaniards are concentrated in the administrative elite and in the merchants and planters group. Less than one-tenth of those in the lower-ranking calidades falls into those two top categories, which suggests that more prestigious occupations may have been effectively reserved for whites.

Table 5.9. Occupational Strata of Adult Males by Calidad, Orizaba, 1791

Occupational Stratum	Española[a,b]	Castizo	Mestizo	Pardo	Unknown	Row Totals
Administrative Elite	249 (53)	3	5	1	0	258
	19.4% (44.5%)	1.8%	0.7%	0.4%		11.0%
Merchants and Planters	241 (53)	6	46	3	0	296
	19.0% (44.5%)	3.6%	6.1%	1.2%		12.6%
Artisans	433 (5)	74	194	44	3	748
	33.7% (4.2%)	45.2%	26.0%	18.3%	75%	32.0%
Laborers	340 (4)	77	468	102	0	887
	26.5% (3.4%)	47.0%	62.6%	42.4%		38.1%
Servants	18 (4)	4	34	12	1	69
	1.4% (3.4%)	2.4%	4.6%	4.7%	25%	2.9%
Slaves	0	0	0	79	0	79
				33.0%		3.4%

Source: AGN Padrones, vol. 19.

[a] Figures in parentheses are the number of European-born Spaniards contained within the larger figures of all españoles.

[b] Percentage figures are by individual rows.

Because the top of the ladder was reserved for whites, one might suspect that men who had climbed to creole status between 1777 and 1791 were perhaps "promoted" in order to match their calidad with the status associated with their occupation. This appears to have been the case with the family of don José Díaz. This creole gentleman had been a mestizo in 1777, as was his wife, doña María Gómez. By 1791, however, they had moved well up the social ladder. Not only was don José now an hacendado, but his two teenage sons were also accorded the honorific *don* in front of their names. One was then a seminarian and the other a student, most likely in Puebla or Mexico City. They also had two teen-age daughters and three more minor children, cared for three more underage nieces and nephews, and lived in a presumably large house on Calle Tercera Real with two mestiza servants.[63]

That a family such as the Díaz clan would be presumed to be on the highest rung of society and that their calidad had risen to match their economic standing come as no surprise. Although David Brading reports that the Mexican secular clergy had come to include castas from artisan families, William B. Taylor emphasizes the importance of creole status to entering the priesthood, as one of the Díaz sons was training to do.[64]

What does emerge from the census manuscript, perhaps unexpectedly, is the discovery that most of those men who moved upward in the sistema de castas between 1777 and 1791 appear to have come from all walks of life. As can be seen in table 5.10, the upward and downward movers are distributed across the occupational groups in proportions similar to those observed for the entire 1791 census, although each of the four men who lost creole status were from the artisan and laborer groups. Even though a high-rank occupation implies creole status, the opposite is untrue, for the bulk of calidad promotions between 1777 and 1791 still occurred in the lower occupational strata.

THE ELITE

No social history is complete without treating all aspects of society, and an analysis of people of lesser status cannot make much sense without a consideration of the elite. An understanding of the society in which lived the

Table 5.10. Occupational Strata of Orizaba Males Whose Calidad Changed on Census Records Between 1777 and 1791

Occupational Stratum	Calidad Category									
	PX	MC	CE	ME	EM	EC	CM	XP	Row	Total
Administrative Elite	0	0	1	3	0	0	0	0	4	(4,0)
Merchants and Planters	0	2	7	10	0	0	2	1	22	(19,3)
Artisans	6	4	16	20	1	1	3	3	54	(46,8)
Laborers	2	5	14	21	1	1	9	0	53	(42,11)
Servants	0	0	0	1	0	0	0	1	2	(1,1)

Source: AGN Padrones, vol. 19; AGI México 2580.
Note: Calidad categories: the first letter in each column is the calidad in the 1777 census manuscript, and the second letter is the calidad in the 1791 manuscript.
E = español, español europeo, hijo de algo; C = castizo; M = mestizo; P = pardo.
PX = a 1777 pardo who changed to any other calidad in the 1791 record. XP = any person whose calidad was changed to pardo in the 1791 census. Figures in parentheses are the subtotals of men who were "promoted" to a more prestigious calidad and those who drifted "downward," respectively.

laborers on the Puente de Escamela similarly requires a study of Orizaba's elite. The first question must therefore be: Who were the elite? A simple question, but one whose answer becomes more complex when one tries to define a dividing line between the elite and the nonelite.

Wealth and power provide a starting point and identify the most obvious members of the elite. Orizaba's prosperity largely came from commercial agriculture. A sugar-growing area throughout much of the colonial period, Orizaba had by 1791 been converted into a center of tobacco production. Susan Deans-Smith has thoroughly analyzed the wealthy planters and important officials of the Royal Tobacco Monopoly, so only a brief sketch of her findings is presented here. Tobacco planters existed across a wide spectrum of status, from poor yndio peasant to rich español hacendado. The wealthiest planters had access to financial credit from the government monopoly and were engaged in a variety of landholding arrangements. Much land around Orizaba belonged to three absentee landlords: the Count of the Valle de Orizaba, the Marquis of the Valle de la Colina, and the Marquis of Sierra Nevada. Many influential planters were of peninsular origin and rented their haciendas and ranchos from one of the three noblemen, whereas many less wealthy creole and peninsular planters owned much smaller plots of land. Other planters rented parts of haciendas, cabildo lands, or land belonging to yndio pueblos.[65]

The social importance of Orizaba's arrendatarios, tenant farmers, stands in contrast to their importance in areas such as Tepeaca, a town near Puebla where most of the elite were creole landowners. Juan Carlos Garaviglia and Juan Carlos Grosso, in their study of Tepeaca, also emphasize the blurred distinctions among planters, merchants, and royal and ecclesiastic officials.[66] We find much the same situation in Orizaba as in Tepeaca and elsewhere in the Spanish Empire. Consider, for example, the case of don Marcos González. Christon Archer introduces González to modern students as a senior militia officer in the Infantry Regiment of Tres Villas in 1799, whereas Susan Deans-Smith highlights his role as a prominent Orizaba planter who rented his tobacco rancho from the Marquis of the Valle de la Colina and who earlier held a commission as captain in the Provincial Regiment of Córdoba.[67] A peninsula-born Spaniard, González was *alcalde ordinario de primer voto* in February 1792, and by 1796 he held the post of *regidor alférez real,* making him senior among the members of the cabildo, and he had already been promoted in the militia to lieutenant colonel.[68] We saw earlier that González played a significant role within the Ayuntamiento de Orizaba during the construction of the Puente de Escamela. González serves as a useful contrast with much poorer creoles who nevertheless might still be considered elite persons.

One indicator of elite status is ascription of the honorific titles *don* and *doña*. In this regard, the Orizaba practice, as reflected in the 1791 census manuscript, departs from trends noted by historians studying other places in Mexico. Only 502 instances of the honorific appear, of which 334 are male and 168 are female. This number equates to one-eighth of the 4,000 españoles and hijos de algo, a sharp contrast with 1821 Guadalajara, where slightly more than one-half the "Spanish" persons were acknowledged as *don* or *doña*.[69] The 1777 census takers in both the Villa de Orizaba and Santa María de Zoquitlán were also more generous: 331 españoles, 363 españolas, 1 castiza, 2 mestizos, and 3 mestizas carried the honorific title. Nevertheless, approximately 1 of every 3 of the 2,132 españoles in the two manuscripts still amounts to a lesser fraction than the figures from Guadalajara. The less frequent use of *don* and *doña* makes it perhaps a more valuable indicator of social elite status than it would in places where the honorific title was more liberally accorded, but even so it still seems to cover a broad spectrum. For instance, at the lower end of the scale lie some petty government officials, who despite their low standing in the bureaucracy still bore the honorific *don*.

Tolls on the Camino Real were collected by guardas de la renta of the Real Aduana, customs officials who briefly appeared in the preceding chapter. One such toll gate was the Garita de Escamela, located close to the bridge upon whose construction much of chapter 4 is centered. Government service records also identify these men and indicate their official salaries. Those guards who enlisted before 1777 received 365 pesos per year. In 1777, the pay rate for customs guards was reduced to 228 pesos and 1 real per year, which equates to 5 reals per day. Thus in 1793, three guards such as don Pedro Barbudo, a forty-five-year-old peninsular Spaniard, received the same daily income as did the stonemasons who labored at the bridge construction site.[70] Of course, the customs guards were paid for every day of the year, and there doubtless existed other perquisites. The 1791 census manuscript suggests that the customs guards lived at the garita itself. The Crown therefore may have provided living quarters at the garita, as did the Royal Tobacco Monopoly for several officials who apparently had apartments within the factory. Most obviously, Barbudo and the other guards may have had opportunities for extralegal personal enrichment.

One cannot easily equate don Pedro Barbudo with don Marcos González, but both stood much higher in society than the artisans and laborers on the Puente de Escamela. Elite people still used their "whiteness" as one important self-descriptor, although whiteness had become less a mark of status because many more nonelite persons also bore that distinction by 1791. Despite the considerable mobility within the sistema de castas and the comparatively large fraction of its nonelite population who claimed status as españoles, Orizaba society still defined itself in terms of social race, and the findings here echo those of Brading and Wu for León, Guanajuato, and Querétaro. Additional research of other places is needed to determine if Orizaba's demonstrated upward mobility within the sistema de castas was unusual or if it typified late-colonial Mexico. Public works such as the Puente de Escamela and the Mexico City–Toluca road existed in many locales in the final decades of colonial rule, which hints at a wider applicability of the trends noted in the surviving expense records.

CONCLUSIONS

However the true nature of colonialism may someday come to be understood, there can be no doubt that it contains negotiated processes of power arrangement. The negotiations can be between groups or individuals, and the same person can be simultaneously an agent of colonialism in one situation and its victim in another. The hacendados between Toluca and Mexico City undoubtedly exercised some form of exploitation of their labor supply while operating their enterprises, but in turn they found themselves unable to avoid contributing sizeable amounts of money for the construction of the highway. Their counterparts in Andalucía succeeded in deflecting the burden. Estates in both cases must have profited from any improvement in the ability to ship their products to market. In both cases, powers higher up made similar demands, but only in the colonial case were those particular demands actually met.

The transition from draft labor to free-wage labor in central Mexico is more than a change in the labor regime. It also marks a change in the nature of the negotiation. No longer would the colonial regime make its demands through brokers such as the governors of indigenous pueblos. Instead, the free-wage system brought the road laborer into direct contact with colonial authorities, the same degree of contact already experienced by artisans. Viceroy Bucareli denied the Ayuntamiento de Puebla's request for a labor draft and told it to get on with bridge construction because that was what a good government should do. When a few years earlier Alcalde Mayor Primo de Rivera argued his case for a change from draft to free-wage labor, he did so on grounds of improved efficiency, not from any moral commitment to freedom or opposition to colonial exploitation of forced labor. In the end, the expanded labor supply that came from a rising and essentially compliant population made the change possible. Falling real wages also corroborate the expanding labor supply. Counterfactual arguments about what might have happened under different historical circumstances are very tenuous indeed. But it is difficult to imagine that a regime willing to twist the arms of its landed elite would shrink from coercing manual labor for road building if it appeared necessary to get people working on the job. One suspects this would be the case in a "noncolonial" situation around Madrid or Toledo as much as it was in the case around colonial Mexico City or Orizaba, and it was clearly the case in other situations around the world that involved corvée labor.

Ethnic or racial distinctions are fundamentally social constructions and frequently present in colonial situations. American Spanish colonial society obviously differed from European Spanish society in the composition of its populations. When applied to the American colonial situation, deep-rooted concerns about limpieza de sangre gave rise to the notion of racial calidad and eventually to the codified sistema de castas. Almost one-third of the thousand persons linked between the two Orizaba census manuscripts had different recorded calidad status from one census to the next. The overwhelming tendency for upward movement testifies to the permeability of the intended barriers. One might possibly see in the figures a collective challenge to this form of stigmatization, but that sort of resistance requires a collective consciousness whose existence is not supported by the evidence. More likely, the calidad promotions reflect renegotiated status by the individuals themselves with the existing social relations of power. In that sense, the promotions actually validated the notion of calidad. Indeed, the perceived value of each individual's heightened status must have depended in some degree on its denial to others.

This social history centers on issues of labor and demography. Comparisons with other secondary works serve as a context for expanded analyses of this study's sources, but clearly much room remains for comparative work in the study of colonialism. Future local studies like this one will help to develop a coherent set of ideas about what colonialism is and was, as well as about what it is and was not. We are a long way from developing a set of concepts with which to compare such apparently disparate situations as first-century Roman colonialism in Judea with Japanese colonialism in twentieth-century Taiwan. Such an edifice of knowledge would be a huge one, but one that can be built only one brick at a time, using microhistories of situations from different locations and times. Overarching syntheses and sweeping macrohistories such as those by Braudel in the end depend on specific studies such as this one, just as the imagery of the impressionist painter Georges Seurat stems from the individual dots of color he placed on the canvas. The search for a macroscopic framework must never lose sight of the individual people themselves. The elite military engineer, the tobacco planter, the day road laborer, the blacksmith's wife—all are historical actors whose agency in microscopic situations creates human history.

APPENDIX

METHODOLOGICAL EXPLANATION

The analyses of labor records in this study are based on lists that name individuals who worked on a particular project. These lists account for the expenditure of government monies and were submitted on a weekly basis. The format was virtually identical throughout, so the lengthy weekly submissions from the Toluca Road construction project were formatted the same as the ones prepared twenty-five years earlier for much less ambitious road repairs around Xalapa. Analysis of these documents follows some of the same principles used to analyze census manuscripts. In particular, I searched out names from one record in other records. It was obvious that not many repeat appearances existed, especially among day laborers. Therefore, I decided that it was safe for these purposes to assume that the appearance of the same name in two records meant that one person had reappeared and that I could discard the other possibility—namely, two persons with same name. In the one case where I could have significantly skewed the results by personal choice, I chose instead to discard the data. Although the format of the records was the same throughout, two of the free-wage labor data sets were incomplete, so it was necessary to handle each set a little differently, as described in the following paragraphs.

In order to analyze the employment patterns of the day laborers who worked on the various projects around Xalapa, I selected sixty-five records. Most of the chosen records were payrolls for each of the four crews over the period November 3, 1767, to February 1, 1768. This was the only three-month period for which a legible, uninterrupted set of payrolls was available for all four crews. To check if the patterns observed in the larger set of documents were also evident at other times, I chose another continuous set of records. This earlier set documents the workers on the crew headed first by don José Ortíz and later by don Manuel Machado. I chose the set dated earlier at random to determine if the trends observed in the larger group of documents were evident in other records as well. Each record documents payments made for one calendar week's work for a particular crew, working under the oversight of one sobrestante, or foreman. The initial impression I received from perusing these records is that day laborers show little work continuity, whereas higher-paid workers often reappear on the subsequent week's payroll.

The validity of an impression can in some measure be tested with an attempt to disprove it. I developed a matrix by name and week from the 877 day-laborer workweeks recorded between November 1767 and February 1768. Most of the day laborers are identified by two given Spanish names, but the names are not followed by a name commonly recognized as a Spanish surname. To construct a work history over the period under consideration, I combined names appearing in different payroll records wherever they were not contradicted by two identically named entries during the same week. For each name on the records, this procedure thus created a putative worker with the maximum possible number of repeat appearances in subsequent weeks. Many ambiguities remained, however, from very common names such as "Salvador de la Cruz," which appears fifteen times. Six of these appearances could be combined to create a putative Salvador de la Cruz, who worked eight of ten consecutive weeks, but because in some cases two laborers in one week had that name, it was obvious that more than one Salvador de la Cruz worked on the construction projects. It would have been easy to create arbitrarily several Salvadors with any number of work patterns, so I decided to remove those cases altogether from the study. Accordingly, I discarded 320 entries, leaving 557 as a sample for analysis.

The records from the year 1767 show that the weekly turnover of personnel remained quite high. Recall from chapter 3 that Alcalde Mayor Primo de Rivera complained chiefly of the ineffectiveness of draft labor and personnel turnover. Because the records from the period around 1757 do not list individual day laborers, a quantitative comparison of turnover remains impossible. From a qualitative standpoint, however, one must suspect that the turnover problem remained far from solved after the change to free-wage labor recruitment for day laborers. This problem can be clearly seen by analyzing the 1767–68 records. Of the 557 entries under consideration, 368 are names that appear only once and are never repeated in any of the fifty-two documents in the sample. Another 67 names appear only in two consecutive weeks, but then never reappear. Thus, at least 435 day laborers worked only one stint of one or two weeks and then left to work elsewhere or possibly to resume their normal agricultural pursuits. This number represents 78.1 percent of the 557 entries under consideration. Four possible stints of three weeks exist, with none longer than that.

A repetition of the process for the smaller group of records dated April through July 1767 indicated that 231 out of the total of 246 workers worked only a one-week stint during that period. Including 3 more workers who appeared only during two consecutive weeks, 84.5 percent of the 277 day-laborer entries on the reports represent individuals who worked only one short period on the project and then left. It should be noted that some persons may have gone to other crews and thus showed longer work histories from April to June, but the results here are roughly consistent with those obtained from scrutiny of the reports from the larger group of four separate road crews. The records analyzed in this study constitute the only continuous series that could be pieced together from the 1767–68 Xalapa documents, and conclusions from such short series would be suspect if both series were not also roughly consistent with the eighteen-month series of weekly payroll records discussed in chapter 3.

The records from Orizaba's Puente de Escamela are a complete set. As in the analysis of the 1766–68 projects around Xalapa, I combined from the records individual line entries with identical names wherever possible. There may thus appear only one Pedro Nolasco when two or more may actually have worked at the Puente de Escamela at different times. Combining the records in this manner allowed me to determine how many persons could have been exceptions to the apparent norm of only one or two short appearances by any day laborer on the Puente de Escamela payrolls. The possibility, of course, remains that this approach may have resulted in a slight undercount, but day laborers whose common names pose this dilemma are so few in number that they do not contradict the trends exhibited by the group as a whole.

Many records from the Toluca Road unfortunately did not survive to be filed at the Archivo General de la Nación in Mexico City. A researcher generally finds a record or two from one period and then a few from another period, and so on. Such scattered data are impossible to analyze, but one set of thirteen continuous weeks' records is on file, and I used these records to draw conclusions from the available data.

Using Microsoft Visual FoxPro 3.0, I created a computer database to analyze the 1777 and 1791 Orizaba and Santa María de Zoquitlán census manuscripts. I established a column for the year 1777 or 1791 and for each data field in the census manuscripts, including two for each person's surnames and two more for each person's given names. In the case of a married woman, I placed the husband's surname in the column for the woman's second surname, followed by "de" to indicate that it was not her own mother's surname. To tabulate occupations in a way that would lend itself to easy sorting, I established a numerical code for each category listed in the 1791 census. I also established additional columns to tabulate data that resulted from my evaluation, such as whether a family structure was nuclear or extended or who was the head of the household at a particular address. Each row entry consisted of one person from one census manuscript and was assigned a sequential number to aid in analyzing the results of sorting the database by a particular column. All of these column headings became blanks on the data entry form, which I filled out with the census information for each enumerated person, with the blanks being arranged in a way to ease the burden of filling out the 15,884 forms. I then used the collected forms to enter the values into the database. After completing the lengthy and tedious work of creating forms and entering their data into the database, I was able to perform analyses and tabulations easily using the sorting tools of the software.

Notes

Chapter One: Introduction

1. Flynn and Giráldez, "Cycles of Silver"; Stein and Stein, *Silver, Trade, and War.*

2. Hareven, *Family Time, Industrial Time.* Important labor-related studies of colonial Mexico include Deans-Smith, *Bureaucrats, Planters, and Workers*; Gibson, *The Aztecs under Spanish Rule*; Haslip-Viera, "The Underclass"; Ladd, *The Making of a Strike*; Patch, *Maya and Spaniard in Yucatan, 1648–1810*; Semo, *Historia del capitalismo en México*; Taylor, *Landlord and Peasant in Colonial Oaxaca*; and Van Young, *Hacienda and Market in Eighteenth-Century Mexico.* Although not directly discussed in this study, the difficulties in studying labor records are suggested by Salvatore and Brown, "Trade and Proletarianization in Late Colonial Banda Oriental," and by Gelman, "New Perspectives on an Old Problem and the Same Source."

Chapter Two: Tentacles of Commerce: *Caminos Reales* in Late Bourbon Mexico

1. Archivo General de Indias (hereafter AGI), Audiencia de México 1447, s/n, letter, Viceroy Azanza to Crown, July 27, 1797.

2. AGI Estado 37, n. 38, letter, Viceroy Azanza to Príncipe de la Paz, June 28, 1799.

3. Valle Pavón, *El camino Mexico-Puebla-Veracruz*, 51–59.

4. AGI México 2506, s/n, letter, Viceroy Revillagigedo to Crown, February 10, 1794.

5. An alcalde mayor was a local magistrate who combined executive and judicial functions, somewhat similar to a town mayor in an English colony. An analogous position, somewhat more prestigious and with more authority, was entitled *corregidor.* After the implementation of administrative reforms in 1786 in New Spain, these officials were superseded by an official entitled *subdelegate,* who was subordinate to a new layer of provincial governor called an *intendant.* For an in-depth treatment of this subject, see Haring, *The Spanish Empire in America.*

6. Archivo General de la Nación (hereafter AGN) Caminos y Calzadas, vol. 5, ff. 380–87, report of don Miguel Cárdenas to Alcade Mayor don Antonio Primo de Rivera, September 7, 1757.

7. Ibid., vol. 8, ff. 238–39, letter from Alcalde Mayor don Vicente de Toledo y Vivero to Viceroy the Marquis of Croix, n.d. In distinguishing between Totonac and Nahua pueblos, I rely on Deans-Smith, "Native Peoples of the Gulf Coast from the Colonial Period to the Present," in particular to the maps on 277–78. On the name "Nahua," see Cline, "Native Peoples of Colonial Central Mexico." Finding a satisfactory catch-all term for the indigenous peoples of Mexico remains unresolved. The term *yndio* appears in this study whenever this juridical term is more appropriate than an ethnic term. This is the spelling of the Spanish translation of the word *Indian* as it appears in the sources of this study. It is used here in an attempt to avoid the pejorative connotations that sadly often attend present-day *indio* and other substitute names.

8. AGN Caminos y Calzadas, vol. 10, ff. 12–13, petition, don Simón Nicolás to Viceroy Martín de Mayorga, n.d., appears to have been written approximately March 26, 1781.

9. A judge of the audiencia, the latter being the supreme court of a large regional division of the empire. Audiencias in the Americas also served an important advisory function for administration, combining executive and judicial functions. See Haring, *The Spanish Empire in America.*

10. AGN Caminos y Calzadas, vol. 10, ff. 15–24, various petitions and order of the oidor fiscal de la Real Audiencia, August 6, 1781.

11. Van Young, *Hacienda and Market in Eighteenth-Century Mexico*, 83.

12. On the matter of indigenous peoples' use of the courts, see especially Borah, *Justice by Insurance*, and Taylor, *Landlord and Peasant in Colonial Oaxaca.*

13. AGN Caminos y Puentes, vol. 2, f. 94, letter, Captain don Miguel del Corral to Viceroy Matías de Gálvez, March 23, 1784. (The spines of volumes 1 through 6 of AGN Caminos y Puentes are actually marked "Fomento de Caminos.")

14. Ibid., vol. 2, ff. 95–96, memorial of Corral, November 17, 1784.

15. AGI México 1238, instructions of Viceroy Revillagigedo to his successor, Viceroy Branciforte, June 30, 1794. Revillagigedo's document outlines the underlying assumptions and what had thus far been achieved.

16. Vinson, *Bearing Arms for His Majesty*, 29.

17. Archer, *The Army in Bourbon Mexico, 1760–1810*, 39–41.

18. Servicio Histórico Militar (hereafter SHM) Colección de Documentos, rollo 62, caja 5-3 to 10-4, "Plan de defensa del Reyno de Nueva España, por las costas colaterales a Vera Cruz dispuesto del orden superior por la Real Junta de Fortificación, celebra en dicha Plaza año de 1774 de orden de Su Majestad," January 10, 1775.

19. Ringrose, "Carting in the Hispanic World."

20. AGN Caminos y Calzadas, vol. 11, f. 53, letter, don Miguel Páez de la Cadena to Viceroy Martín de Mayorga, October 20, 1779, with annotated endorsement.

21. Ibid., vol. 11, f. 54, letter, Juan Navarro to Viceroy Matías de Gálvez, June 28, 1783.

22. Rivera Cambas notes that the fortress of San Carlos de Perote mounted fifty-nine artillery pieces. See his *Historia Antigua y Moderna de Xalapa*, 1: 164–65.

23. Many works deal with these famines, but see especially Florescano, *Precios del maíz y crisis agrícolas en México (1708–1810)*, 124–25; Ouweneel, *Shadows over Anáhuac*, 109–12; and Tutino, *From Insurrection to Revolution in Mexico*, 74.

24. AGI México 1238, item 163, letter, Viceroy Revillagigedo to his successor, Viceroy Branciforte, June 30, 1794.

25. Kanter, *"Hijos del pueblo,"* 42–44.

26. AGN Caminos y Calzadas, vol. 22, f. 257, letter, don Bernardino Bonavía to Viceroy Revillagigedo, April 8, 1791. On Bonavía's offices, see Arnold, *Bureaucrats and Bureaucracy in Mexico City*, 37.

27. AGN Caminos y Calzadas, vol. 11, ff. 281–83, letter, Captain don Manuel Agustín Mascaró to Viceroy Revillagigedo, May 8, 1791.

28. Ibid., vol. 11, ff. 289–301, letter, Captain Mascaró to Viceroy Revillagigedo, June 4, 1791.

29. Ibid., vol. 11, f. 341, and vol. 13, ff. 49–231, various memorials and letters regarding the financing of the project.

30. Ibid., vol. 11, ff. 267–78, letter, don José Sanz to Viceroy Revillagigedo, May 16, 1791. This document also promulgates the tolls to be charged to various travelers.

31. Ibid., vol. 11, f. 300, letter, Captain Mascaró to Viceroy Revillagigedo, June 4, 1791.

32. Ibid., vol. 11, ff. 316–23, letter, don Tomás Hidalgo, *escribano real*, to Viceroy Revillagigedo, April 13, 1792.

33. Ibid., vol. 11, ff. 356–58, letter, Viceroy Revillagigedo to Crown, February 4, 1793.

34. Ibid., vol. 13, f. 151, letter, Captain Mascaró to Viceroy Revillagigedo, October 17, 1793, itemizes the tool shortages that needed to be resolved before actual work could commence.

35. Díaz-Trechuelo Spinola, Pajarón Parody, and Rubio Gil, "El Virrey don Juan Vicente de Güemes," 134.

36. Archivo General de Simancas (hereafter AGS) Secretaría de Guerra 7241, exp. 51, ff. 239–41, documents contained in Mascaró's personnel record.

37. Jurado Sánchez, *Los caminos de Andalucía*, 97–102, and Uriol Salcedo, *Historia de los caminos de España*, 1: 242, 257–63.

38. Valle Pavón, "El consulado de comerciantes de la ciudad de México y las finanzas novohispanas"; Borchart de Moreno, *Los mercaderes y el capitalismo en la ciudad de México*; Kizca, *Colonial Entrepeneurs*; see also Brading, *Miners and Merchants in Bourbon Mexico*.

39. AGN Caminos y Calzadas, vol. 13, f. 258, letter, Viceroy Revillagigedo to Captain Mascaró, February 5, 1793.

40. Souto Mantecón, *Mar abierto*, 49, 55–71. See also Booker, *Veracruz Merchants, 1770–1829*, 41–60.

41. Although not written until January 9, 1800, "Memoria sobre la necesidad y utilidades de la construcción de un camino carretero desde Veracruz a México" of the Consulado de Veracruz well summarizes the Veracruz merchants' position and appears in its entirety in Ortíz de la Tabla Ducasse, *Memorias políticas y ecónomicas del Consulado de Veracruz, 1796–1822*, 27–44. The original document is in AGI México 2996.

42. Souto Mantecón, *Mar abierto*, 71–86, and Booker, *Veracruz Merchants*, 50–52.

43. Walker's *Spanish Politics and Imperial Trade, 1700–1789*, analyzes at length the Xalapa trade fairs.

44. Antolín Espino and Navarro García, "El Virrey Marqués de Branciforte," 371, 374.

45. AGI Estado 25, n. 42, letter, Viceroy Branciforte to Príncipe de la Paz, September 26, 1796. Also see Suárez Argüello, *Camino real y carrera larga*.

46. AGI Estado 25, n. 42, February 11, 1797, endorsement of Príncipe de la Paz on Viceroy Branciforte letter of September 26, 1796.

47. Ibid., n. 41, letter, Viceroy Branciforte to Príncipe de la Paz, September 27, 1796.

48. Ibid., n. 59, "Testimonio del expediente formado sobre apertura de un camino recto desde Toluca hasta Celaya" of the consulado, signed by the Count of Contramina and don Antonio Bassoco, October 5, 1796.

49. Ibid., n. 59, "Testimonio de las últimas actuaciones en el expediente formado sobre apertura de un camino recto desde esta capital a Veracruz por Puebla y las villas de Orizaba y Córdoba," n.d.

50. Ibid., n. 59, petition of the pueblo Ameca to Viceroy Branciforte, October 28, 1796.

51. Perhaps most entertainingly described in an 1891 novel by Manuel Payno, *Los bandidos de Río Frío*, but see also Payno's account of his 1843 stagecoach journey from Mexico City to Veracruz, *Un viaje á Veracruz en el invierno de 1843*.

52. AGI Estado 25, n. 59, letter, Viceroy Branciforte to Príncipe de la Paz, October 27, 1796.

53. Antolín Espino and Navarro García, "El Virrey Marqués de Branciforte," 508.

54. AGI Estado 25, n. 92, proclamation of Lic. Don Luis Gonzaga de Ybarrola, Abogado de la Real Audiencia, December 9, 1796.

55. Ibid., n. 92, letter, Viceroy Branciforte to Príncipe de la Paz, December 29, 1796.

56. Ibid., n. 104, letter, Viceroy Branciforte to Príncipe de la Paz, December 28, 1796.

57. Souto Mantecón, *Mar abierto*, 120.

58. Ibid., n. 41, letter, don Pedro de Mantilla to Príncipe de la Paz, February 3, 1797.

59. AGI Estado 25, n. 59, letter, Crown to Viceroy Branciforte, February 11, 1797.

60. Ibid., n. 42, letter, Consulado de México to Viceroy Azanza, August 31, 1798.

61. AGN Caminos y Calzadas, vol. 18, f. 52, letter, Ayuntamiento de Orizaba to Viceroy Branciforte, August 18, 1797; and vol. 18, ff. 56–57, letter, Ayuntamiento de Orizaba to Viceroy Branciforte, September 23, 1797.

62. Ibid., vol. 18, f. 61, Order of the Intendant of Veracruz, October 16, 1797.

63. AGI Estado 37, n. 38, letter, Consulado de México to Viceroy Azanza, June 28, 1799.

64. AGI Estado 25, n. 42, letter, Viceroy Azanza to Príncipe de la Paz, June 28, 1799.

65. Brading, *Miners and Merchants*, 124–28.

66. Consulado de Veracruz, "Memoria sobre . . . un camino carretero," in Ortíz de la Tabla Ducasse, *Memorias políticas*, 27–44.

67. Ibid., 39.

68. Souto Mantecón, *Mar abierto*, 97–132.

69. Archivo Municipal de Orizaba (hereafter AMO), caja 9, letter, Consulado de México to Ayuntamiento de Orizaba, November 10, 1802.

70. Souto Mantecón, *Mar abierto*, passim, and Ortíz de la Tabla Ducasse, "Introducción," in *Memorias políticas*, xix.

71. Ortíz de la Tabla Ducasse, *Comercio exterior de Veracruz, 1778–1821*, 78.

72. AGN Caminos y Puentes, vol. 1, ff. 4–10, toll regulation of the Garita de San Martín, February 28, 1804.

73. AGI Estado 25, n. 87, letter, Consulado de México to Viceroy José de Iturrigaray, February 28, 1804.

74. Ibid., n. 87, letter, Viceroy Iturrigaray to Consulado de México, March 27, 1804.

75. See, for example, Lockhart, *The Nahuas after the Conquest*, 163–77.

76. AGI Estado 30, n. 87, letter, José Vicente de Olloquí to Viceroy Iturrigaray, April 22, 1807.

77. Ibid., n. 87, letter, Consulado de México to Viceroy Iturrigaray, June 19, 1807; letter, Iturrigaray to Real Tribunal del Consulado de México, June 30, 1807.

78. Ibid., n. 87, letter, Consulado de México to Viceroy Iturrigaray, June 19, 1807.

79. Ibid., n. 87, letter, Viceroy Iturrigaray to Crown, July 16, 1807.

80. Arróniz, *Ensayo de una historia de Orizaba*, 2: 228.

81. For two classic accounts, see Payno, *Un viaje á Veracruz en el invierno de 1843*, and Calderón de la Barca, *Life in Mexico during a Residence of Two Years in That Country*, 24–31.

82. Coatsworth, "Obstacles to Economic Growth in Nineteenth-Century Mexico," 94.

83. AGN Caminos y Puentes, vol. 1, ff. 1–3, "Noticias de las obras efectuadas en el camino de Veracruz por Jalapa," n.d.

84. Ortíz de la Tabla Ducasse, "Introducción," in *Memorias políticas*, xxvii.

85. Rees, *Transportes y comercio entre México y Veracruz, 1519–1910*, 80.

86. Souto Mantecón, *Mar abierto*, 114.

87. Stein and Stein, *The Colonial Heritage of Latin America*.

88. For a non-Marxist critique of dependency, see especially Haber, "Introduction: Economic Growth and Latin American Historiography"; for a more orthodox Marxist critique, see Jones, "'Business Imperialism' and Argentina 1875–1900"; and for a case study, see Gootenberg, *Between Silver and Guano*.

89. Patch, *Maya and Spaniard in Yucatan*, and Van Young, *Hacienda and Market in Eighteenth-Century Mexico*. In an opposing interpretation, Suárez Argüello contends that the colonial Mexican economy was indeed far more integrated than Patch and Van Young believe. See Suárez Argüello, *Camino real y carrera larga*.

90. See Brading, *Miners and Merchants*; Brading and Cross, "Colonial Silver Mining: Mexico and Peru"; Brading, "Mexican Silver-Mining in the Eighteenth Century"; Bakewell, *Silver Mining and Society in Colonial Mexico*; and Ladd, *The Making of a Strike*.

91. Garner, *Economic Growth and Change in Bourbon Mexico*, and Rees, *Transportes y comercio entre México y Veracruz, 1519–1910*, 16.

CHAPTER THREE: FROM DRAFT TO FREE-WAGE LABOR: PROJECTS AROUND XALAPA DURING THE MID-EIGHTEENTH CENTURY

1. AGN Caminos y Calzadas, vol. 5, ff. 278–79, letter, Primo de Rivera to Ahumada, June 21, 1758.

2. Ibid., vol. 1, ff. 114–19, petitions and letters relating to the wooden bridge at Coatepec, August 2 to August 8, 1736.

3. Ibid., vol. 5, f. 35, letter, Alcalde Mayor don Antonio Primo de Rivera to Viceroy don Agustín Ahumada, October 25, 1757.

4. Ibid., vol. 5, ff. 136–38, report of Primo de Rivera, March 2, 1759.

5. Ibid., vol. 5, ff. 323–51, report of Primo de Rivera, February 14, 1758.

6. Van Young, *La crisis del orden colonial*, 94.

7. Martin, *Governance and Society in Colonial Mexico*, 100–102, 106–7, 120–24.

8. AGN Caminos y Calzadas, vol. 5, ff. 136–38, 323–51.

9. AGI México, legajo 2580, Padrón General de Jilotepec, 1777.

10. Cook and Borah, *Essays in Population History: Mexico and the Caribbean*, 1: 286–90.

11. Baskes, *Indians, Merchants, and Markets*, 34.

12. AGN Caminos y Calzadas, vol. 5, ff. 136–38, 323–31.

13. Gibson, *The Aztecs under Spanish Rule*, 209.

14. Vaporis, "Post Station and Assisting Villages: Corvée Labor and Peasant Contention."

15. AGN Caminos y Calzadas, vol. 5, ff. 323–51.

16. Along with other related documents, the reports are contained in ibid., vol. 8, ff. 151–354.

17. Ibid., vol. 5, f. 13, report of Antonio de Mirón, April 14, 1767.

18. Haskett, "Living in Two Worlds," 42, 52–54, and Haskett, *Indigenous Rulers*, 99–104.

19. Stern, *The Secret History of Gender*, 194–95.

20. Escobar Ohmstede, "Del gobierno indígena al ayuntamiento constitucional en las Huastecas hidalguense y veracruzana, 1780–1853."

21. Lockhart, *The Nahuas after the Conquest*, 55; Terraciano, *The Mixtecs of Colonial Oaxaca*, 197; Patch, *Maya and Spaniard in Yucatan*, 24–26; Farriss, *Maya Society under Colonial Rule*, 232–36; Restall, *The Maya World*, 62–65; Spalding, *Huarochirí*, 291–93; and Stern, "The Age of Andean Insurrection, 1742–1782," 71–76.

22. Gibson, *The Aztecs under Spanish Rule*, 194–219.

23. McWatters, "The Royal Tobacco Monopoly in Bourbon Mexico," 201, 216, and Deans-Smith, *Bureaucrats, Planters, and Workers*.

24. Davis, *Society and Culture in Early Modern France*, 71.

25. See in particular the essays in Schroeder, Wood, and Haskett, eds., *Indian Women of Early Mexico*.

26. Deans-Smith, *Bureaucrats, Planters, and Workers*, 122, 295, based on AGN Tabaco, vol. 352.

27. AGN Obras Públicas, vol. 8, ff. 1–7.

28. Haskett, *Indigenous Rulers*, 101.

29. Conrad and Meyer, "The Economics of Slavery in the Ante Bellum South."

30. Martin, *Governance and Society in Colonial Mexico*, 56–59.

31. Van Young, *Hacienda and Market in Eighteenth-Century Mexico*, 245–64.

32. Taylor, *Landlord and Peasant in Colonial Oaxaca*, 149.

33. Gibson, *The Aztecs under Spanish Rule*, 251.

34. Van Young, *La crisis del orden colonial*, 85.

35. Ibid., 91.

36. Martin, *Governance and Society in Colonial Mexico*, 97–124.

37. AGN Caminos y Calzadas, vol. 8, ff. 153–56, payment reports for road construction between Xalapa and Perote, April 27 to May 18, 1767.

38. Ibid., vol. 8, ff. 287, 251, 362, 324, payment reports, December 28, 1767.

39. Hindley, *A History of Roads*, 54–56.

40. Brockington, *The Leverage of Labor*, 167.

41. AGN Caminos y Calzadas, vol. 15, ff. 380–81, report of Miguel Cárdenas to Alcalde Mayor don Antonio Primo de Rivera, September 7, 1757.

42. Semo, *Historia del capitalismo en México*, 58, 189, 251–52.

43. Wolf, *Europe and the People without History*, 79–88.

44. Amin, *Class and Nation*, x, 2.

45. Bobb, *The Viceregency of Antonio María Bucareli in New Spain, 1771–1779*, 3–32, 259–70. See also Herr, *The Eighteenth Century Revolution in Spain*, 52; Lynch, *Bourbon Spain 1700–1808*, 256; and Engstrand, "The Enlightenment in Spain: Influences upon New World Policy," 436.

46. Brunt, *Italian Manpower, 225 BC–AC 14*, 199; Chevallier, *Roman Roads*, 84, 156; Von Hagen, *The Roads That Led to Rome*, 33–36, 67–70; Casson, *Travel in the Ancient World*, 164, 166, 172; Keay, *Roman Spain*, 88.

47. Inalcik, *The Ottoman Empire: The Classical Age 1300–1600*, 146–49.

48. Shaw, *History of the Ottoman Empire*, 1: 128–29, 136, 151, 161.

49. Joyce and Winter, "Ideology, Power, and Urban Society in Pre-Hispanic Oaxaca"; Kowalewski and Finsten, "The Economic System of Ancient Oaxaca," 421–22; Carrasco, "The Political Economy of the Aztec and Inca States"; and LeVine, "Inka Labor Service at the Regional Level."

50. Murra, *The Economic Organization of the Inka State*; D'Altroy and Earle, "Staple Finance, Wealth Finance, and Storage in the Inka Political Economy"; and LeVine, "Inka Labor Service."

51. Waldron, "The Problem of the Great Wall of China," 655.

52. Chan, "The Organization and Utilization of Labor Service under the Jurchen Chin Dynasty," 614, and Ebrey, "The Economic and Social History of Later Han," 613–14.

53. Wright, "The Sui Dynasty (581–617)," 135.

54. Chan, "Labor Service under the Jurchen Chin," 615.

55. Brook, "Communications and Commerce," 601–9.

56. Myers and Wang, "Economic Developments, 1644–1800," 593–95.

57. Patch, *Maya and Spaniard in Yucatan*, 147.

58. Restall, *The Black Middle*, chap. 3.

59. McCreery, *Rural Guatemala, 1760–1940*, 2–6, 303–7.

60. Stein, "Myth and Ideology in a Nineteenth Century Peruvian Peasant Uprising," 238; Newbury, "Colonialism, Ethnicity, and Rural Political Protest," 263; and Brocheux, "Moral Economy or Political Economy?" 796–97.

61. Turits, *Foundations of Despotism*, 106, 113.

62. Chevalier, *Land and Society in Colonial Mexico*. And among many studies of the hacienda, see especially Gibson, *The Aztecs under Spanish Rule*; Patch, *Maya and Spaniard in Yucatan*; Taylor, *Landlord and Peasant in Colonial Oaxaca*; and Van Young, *Hacienda and Market in Eighteenth-Century Mexico*.

63. Stern, "Feudalism, Capitalism, and the World-System in the Perspective of Latin America and the Caribbean," 865–72.

64. The matter of population evokes here the issues of the so-called Brenner Debate, in which Robert Brenner challenged neo-Malthusian arguments about the centrality of demographic change in the emergence of European capitalism. The poles of the debate were set by Emmanuel Le Roy Ladurie, *The Peasants of Languedoc*, and Robert Brenner, "Agrarian Class Structure and Economic Development in Pre-industrial Europe." Brenner later expanded his views in "The Agrarian Roots of European Capitalism." His articles are reprinted in Aston and Philpin, eds., *The Brenner Debate: Agrarian Class Structure and Economic Development in Pre-industrial Europe*. Although demographic matters are crucially important to this study, the Brenner Debate itself is irrelevant here because I contend that corvée labor systems are not associated with any particular mode of production.

CHAPTER FOUR: TIMES GET TOUGHER: THE PUENTE DE ESCAMELA AND THE TOLUCA ROAD

1. For a description written a few years after the period studied here, see von Humboldt, *Political Essay on the Kingdom of New Spain*, 2: 250–70.

2. In particular, see Arróniz, *Ensayo de una historia de Orizaba*; Deans-Smith, *Bureaucrats, Planters, and Workers*; Carroll, *Blacks in Colonial Veracruz*; and McWatters, "The Royal Tobacco Monopoly in Bourbon Mexico, 1764–1810."

3. AGN Clero Regular y Secular, vol. 51, ff. 56–59, letter, Doctor don Francisco Antonio de Yllueca, Cura Vicario y Juez Eclesiástico del Partido de Orizaba, to the Bishop of Puebla, dated July 18, 1770.

4. AGN Caminos y Puentes, vol. 1, f. 99, report of Captain don Miguel del Corral to Viceroy Matías de Gálvez, August 17, 1784.

5. AGN Peajes, vol. 1, f. 48, order of Viceroy Revillagigedo dated March 16, 1791.

6. Arróniz, *Ensayo de una historia de Orizaba*, 226.

7. AGN Peajes, vol. 1, ff. 52–195. The series contains no report for the week of January 22, 1792, and contains two separate reports for the weeks of April 30 and May 8,

1791, during which some laborers quarried stone at a location remote from the construction site itself.

8. The Revillagigedo census manuscript for Orizaba is contained in AGN Padrones, vol. 19. For a discussion of this census in general, see Cook and Borah, *Essays in Population History: Mexico and the Caribbean*, 1: 44.

9. Deans-Smith, "Native Peoples of the Gulf Coast from the Colonial Period to the Present," 276.

10. Ajofrín, *Diario del viaje*, 110.

11. AGN Padrones, vol. 19, f. 372, and AGN Peajes, vol. 1, ff. 52–195.

12. AGN Padrones, vol. 19, f. 254, and AGN Peajes, vol. 1, ff. 82–93.

13. Ortíz, *Cuban Counterpoint*, 56. Also see Deans-Smith, *Bureaucrats, Planters, and Workers*, 293 n. 58.

14. AGN Peajes, vol. 1, ff. 67–173.

15. AGN Padrones, vol. 19, f. 372.

16. Ibid., vol. 19, f. 241.

17. AGN Peajes, vol. 1, ff. 57–59.

18. Ibid., vol. 1, f. 64.

19. Ibid., vol. 1, ff. 51–54.

20. Ibid., vol. 1, ff. 163, 166; AGN Obras Públicas, vol. 39, f. 101; and AGN Padrones, vol. 19, f. 137.

21. Deans-Smith, *Bureaucrats, Planters, and Workers*, 208–9.

22. AGN Padrones, vol. 19, ff. 261, 269, and AGN Peajes, vol. 1, ff. 62, 67, 69, 76, 79, 180.

23. Van Young, *La crisis del orden colonial*, 95–99.

24. Díaz-Trechuelo Spinola, Pajarón Parody, and Rubio Gil, "El Virrey don Juan Vicente de Güemes," 134.

25. AGN Caminos y Calzadas, vol. 14, ff. 184–360, Brigada del Poniente payroll records from August 4 to October 31, 1794, and vol. 13, f. 281, letter, Captain don Diego García Conde, dated February 20, 1794.

26. Florescano, *Precios del maíz y crisis agrícolas en México*, 92–95, and Gibson, *The Aztecs under Spanish Rule*, 308.

27. For an analysis of life and labor in maize production around Guadalajara, which would likely have applied to Toluca as well, see Van Young, *Hacienda and Market in Eighteenth-Century Mexico*, 236–64.

28. AGN Caminos y Calzadas, vol. 13, f. 262, letter, subdelegate to viceroy, November 12, 1793.

29. Ibid., vol. 13, f. 267, letter, Captain don Manuel Agustín Mascaró to viceroy, December 15, 1793.

30. Ibid., vol. 14, ff. 213–18, reports of the Cuadrilla de la Ciudad de Toluca, August 4 to August 19, 1794.

31. Santamaría, *Diccionario de mejicanismos*.

32. Kanter, "Hijos del pueblo," 41, 54.

33. Haslip-Viera, "The Underclass," 294–95.

34. Deans-Smith, *Bureaucrats, Planters, and Workers*, 118, and Van Young, *La crisis del orden colonial*, 35.

35. AGN Peajes, vol. 1, ff. 113–18.

36. Van Young, *La crisis del orden colonial*, 95.

37. Haslip-Viera, "The Underclass," 295.

38. Garner, *Economic Growth and Change in Bourbon Mexico*, 33.

39. AGN Tabaco, vol. 527, report of don Miguel Costansó, dated June 1, 1793.

40. Ladd, *The Making of a Strike*.

41. Bakewell, *Silver Mining and Society in Colonial Mexico*, 126, and Brading and Cross, "Colonial Silver Mining: Mexico and Peru," 558.

42. Brockington, *The Leverage of Labor*, 166–68; Gibson, *The Aztecs under Spanish Rule*, 137–48; and Cook and Borah, *Essays in Population History: Mexico and the Caribbean*, 1: 319–21.

43. AGN Caminos y Calzadas, vol. 14, ff. 213–16, reports of the Cuadrilla de la Ciudad de Toluca, August 4 to August 25, 1794.

44. Ibid., vol. 6, f. 1, report of Lieutenant Colonel don Diego García Conde to Viceroy José Iturrigaray, February 1, 1804.

45. AGN Peajes, vol. 1, ff. 82, 96, 145, 154.

46. Ibid., vol. 1, ff. 113, 118.

47. Ibid., vol. 1, ff. 154–56, 166.

48. Cook and Borah, *Essays in Population History: Mexico and the Caribbean*, 1: 255.

49. See, for example, Carroll, *Blacks in Colonial Veracruz*.

50. AGN Padrones, vol. 19, f. 235. Ages of heads of household normally were recorded in this census, and the omission of Estevan's age appears to have been an oversight.

51. Haslip-Viera, "The Underclass," 294.

52. AGN Padrones, vol. 19, f. 34.

53. AGN Peajes, vol. 1, ff. 122–45.

54. Ibid., vol. 1, f. 124.

55. Thus far I have been able to locate only one document confirming the existence of a carpenters' guild in Orizaba. In a letter to the Ayuntamiento de Orizaba dated November 8, 1803, gremio officials requested permission to establish a Monte de Piedad like those that other guilds operated. This guild's existence most likely substantially predated both the surviving letter and the 1791–92 construction project as well. AMO, caja 10.

56. AGN Caminos y Calzadas, vol. 18, ff. 279–80, and Haslip-Viera, "The Underclass," 295.

57. Garner, *Economic Growth and Change in Bourbon Mexico*, 32, 35.

58. AGN Caminos y Calzadas, vol. 14, f. 301, Brigada del Levante pay report of September 1 to September 6, 1794.

59. Ibid., vol. 14, ff. 184–360.

60. Ibid., vol. 14, ff. 356–54, Brigada del Poniente reports from October 4 to October 18, 1794.

61. Ladd, *The Making of a Strike*, 96; Florescano Mayet, *El camino México-Veracruz en la época colonial*, 80; and Hamnett, *Roots of Insurgency: Mexican Regions, 1750–1824*, 85.

62. On the demographic changes in the eighteenth century, see Cook and Borah, *Essays in Population History: Mexico and the Caribbean*, 1: 89–102, 320. For the increased use of free-wage labor in both urban and rural situations, see, among others, Brading, *Haciendas and Ranchos in the Mexican Bajío*, 176–77; Tutino, *From Insurrection to Revolution in Mexico*, 69–70; and Van Young, *Hacienda and Market*, 268.

63. Haslip-Viera, "The Underclass," 295.

64. Thompson, *The Making of the English Working Class*, 215, 235.

65. Ladd, *The Making of a Strike*, 90–97; Brading, "Mexican Silver-Mining in the Eighteenth Century," 680; and Matthew Restall, personal communication to author, August 1, 2003.

66. In addition to Garner, *Economic Growth and Change in Bourbon Mexico*, see Van Young, *The Other Rebellion*.

67. Garner, *Economic Growth and Change in Bourbon Mexico*, 33.

68. AGI Indiferente General 184, *hojas de servicio* of various officials for the years 1793, 1798, and 1808.

69. AGI Indiferente General 175, ex1, ff. 1–15, documents relating to the employees of the Royal Tobacco Monopoly.

70. Ibid., ex1, f. 3; AGN Padrones, vol. 19, f. 113.

71. Archer, *The Army in Bourbon Mexico*, 194.

72. AGI Indiferente General 1906, f. 1411, no date.

73. AGS Secretaría de Guerra 7241, exp. 1/1/17. Military service record of Captain Manuel Agustín Mascaró, December 31, 1793. See also numerous references in González Tascón, *Ingeniería española en el Ultramar*.

74. Many diary accounts survive, including Calderón de la Barca, *Life in Mexico*, but see especially the excerpted accounts appearing in several secondary histories of the War of 1847: Eisenhower, *Agent of Destiny: The Life and Times of General Winfield Scott* and *So Far from God: The U.S. War with Mexico, 1846–1848*; Bauer, *The Mexican War*; and Henry, *The Story of the Mexican War*.

75. AGI México 1447, s/n, letter, Viceroy Azanza to Crown, July 27, 1798.

76. AGN Tabaco, vol. 307, f. 27, report of Rafael García, inspector of the Royal Tobacco Factory, April 6, 1805, ff. 57–58, memorandum of Real Tribunal del Consulado, May 5, 1805.

77. AGN Obras Públicas, vol. 6, ff. 337–44, letter, don José Díaz Camacho, administrator of haciendas San Bartolomé and Rincón del Rinal, to Subdelegate of Orizaba don Luis José de Segovia, May 18, 1807, and letter, Manuel Rojas, road work director, to subdelegate, September 22, 1807.

CHAPTER FIVE: PATTERNS OF PEOPLE'S LIVES: SOCIETY AND FAMILY IN ORIZABA, 1777–1791

1. AGN Subdelegados, caja 10, ff. 391–406, contains the various letters documenting this incident.

2. AGN Padrones, vol. 19, dated November 16, 1791.

3. AGI México 2309, exp. s/n, ff. 1–19, "Condiciones de la contrata celebrada con los cosecheros de la villa de Orizaba," 1781.

4. AGN Padrones volume 19 is the 1791 Orizaba manuscript, and volume 20 contains the parallel document from Xalapa. The census taken under church auspices at the behest of Viceroy Bucareli is in AGI México 2580.

5. AGI México 2580, f. 398.

6. My choice of 0.95 is somewhat arbitrary, being the lower limit of parity (1.00) plus or minus 5 percent. The 5 percent figure comes from two scholars' contention that a normal ratio of males to females at baptism is 1.05; see Willigan and Lynch, *Sources and Methods of Historical Demography*, 65.

7. Rabell Romero, "Estructuras de la población y características de los jefes de los groupos domésticos en la ciudad de Antequera (Oaxaca), 1777," 284, and Minchom, *The People of Quito, 1690–1810*, 146.

8. See, in particular, Lynch, *Bourbon Spain, 1700–1808*, 374–421.

9. Archer, *The Army in Bourbon Mexico*, 225–26.

10. Brading, "Grupos étnicos, clases, y estructura ocupacional en Guanajuato, 1792," 461.

11. I prepared a computerized database from both sets of manuscripts. The database consists of thirty-nine columns serving as data fields for 15,884 row entries. The rows are for individuals, with a few exceptions, such as the priests of the Carmelite convent or the five hundred unnamed children noted earlier. The column headings consist of categories directly from the census manuscript, such as surname or calidad, and information I assigned to facilitate sorting and searching.

12. Brading and Wu, "Population Growth and Crisis: León, 1720–1860," 2.

13. Deans-Smith, *Bureaucrats, Planters, and Workers*, 109–10.

14. Chance and Taylor, "Estate and Class in a Colonial City"; and McCaa, Schwartz, and Grubessich, "Race and Class in Colonial Latin America," with reply by Chance and Taylor. These articles established the poles of the debate. A number of works cited elsewhere also constitute part of this historiography, but on the matter in general one should see: Strauss, "Measuring Endogamy," 225–45; McCaa, "Modeling Social Interaction"; Seed and Rust, "Estate and Class in Colonial Oaxaca Revisited"; McCaa and Schwartz, "Measuring Marriage Patterns"; Rust and Seed, "Equality of Endogamy"; Arrom, "Marriage Patterns in Mexico City, 1811"; Arrom, *The Women of Mexico City, 1790–1857*; Brading, "Grupos étnicos"; Brading and Wu, "Population Growth and Crisis"; Wu, "The Population of the City of Querétaro in 1791"; Seed, "Social Dimensions of Race: Mexico City, 1753;" and Love, "Marriage Patterns of Persons of African Descent in a Colonial Mexico City Parish." Data from the Sagrario Metropolitano, a parish in the center of late-

seventeenth-century Mexico City, can be found in Cope, *The Limits of Racial Domination*, and data for an early-nineteenth-century location can be found in Anderson, "Race and Social Stratification."

15. On the conversion from sugar to tobacco production, resulting from actions of the Royal Tobacco Monopoly, see Deans-Smith, *Bureaucrats, Planters, and Workers*, 112–17, and McWatters, "The Royal Tobacco Monopoly in Bourbon Mexico," 71–79.

16. Klor de Alva, "*Mestizaje* from New Spain to Aztlán," 59–60. On the sistema de castas generally, see Mörner, *Race Mixture in the History of Latin America*.

17. See, for example, the series of casta paintings in Farmer and Katzew, eds., *New World Orders*, plates 30–45.

18. AGN Padrones, vol. 19, f. 160. Determining the precise meaning of the terms *labrador* and *jornalero* from dictionaries is difficult, and in the absence of Mexican sources I turned to studies of Spain in an attempt to clarify the definitions. For *labrador*, see Vassberg, *The Village and the Outside World in Golden Age Castile*, 58, 86–87. On the meaning of *jornalero*, see Lynch, *Bourbon Spain*, 199. On the issue of land ownership in Orizaba, see Deans-Smith, *Bureaucrats, Planters, and Workers*, 111.

19. AGN Peajes, vol. 1, ff. 122, 173.

20. AGI México 2580, f. 130.

21. AGN Padrones, vol. 19, f. 233.

22. For example, see Chance and Taylor, "Estate and Class in a Colonial City."

23. Degler, *Neither Black nor White*.

24. See Farmer and Katzew, eds., *New World Orders*, 21, figure 19, for a relevant casta painting.

25. AGN Padrones, vol. 19, f. 246.

26. AGN Peajes, vol. 1, ff. 118, 122, 125, 131, 132.

27. AGN Padrones, vol. 19, f. 73, and AGI México 2580, f. 210.

28. AGN Padrones, vol. 19, f. 282, and AGI México 2580, f. 191.

29. Carroll, *Blacks in Colonial Veracruz*, 124–29.

30. AGN Padrones, vol. 20.

31. Chance, *Race and Class in Colonial Oaxaca*, 157.

32. Aguirre Beltrán, *La población negra de México, 1519–1810*, 169.

33. AGN Padrones, vol. 19, ff. 396–97, and AGI México 2580, f. 102.

34. Vinson, *Bearing Arms for His Majesty*, 57–61.

35. Cope, *The Limits of Racial Domination*, 106–24.

36. Chance and Taylor, "Estate and Class in a Colonial City," 466.

37. Garaviglia and Grosso, "El comportamiento demográfico de una parroquía poblana," 627–33; Taylor, *Drinking, Homicide, and Rebellion*, 25; and Ouweneel, *Shadows over Anáhuac*, 14–15. For a discussion of similar findings for late-seventeenth-century Mexico City, see Cope, *The Limits of Racial Domination*, 55.

38. AGN Padrones, vol. 19, ff. 88–89, 167, and AGI México 2580, ff. 167, 194.

39. McCaa, "*Calidad, Clase*, and Marriage in Colonial Mexico: The Case of Parral," 497–501.

40. Chance, *Race and Class in Colonial Oaxaca*, 176–79.

41. Twinam, *Public Lives, Private Secrets*, 54–55, 289–90, 356.

42. Tilly, "Retrieving European Lives," 27–29.

43. Cope, *The Limits of Racial Domination*, 109–21.

44. AGN Padrones, vol. 19, ff. 3, 44.

45. Socolow, *The Merchants of Buenos Aires, 1776–1810*, and "Marriage, Birth, and Inheritance."

46. AGN Padrones, vol. 19, ff. 164, 165.

47. For definitions used in this study, see Ruggles, *Prolonged Connections*, 208–14.

48. Calero, *Orizaba*, 83.

49. On the definition of *operario*, I rely here on McWatters, "The Royal Tobacco Monopoly in Bourbon Mexico," 237. Some confusion arises from the appearance in the 1791 census manuscript of *operario* as the occupation of a number of persons who lived on outlying ranchos, so it seems unlikely that those persons would have worked at the Orizaba factory making cigarettes.

50. AGN Peajes, vol. 1, ff. 118, 162–75, and a copy of the report of January 15, 1792, which is missing from Peajes, vol. 1, is in AGN Obras Públicas, vol. 39, f. 101.

51. Anderson, "Race and Social Stratification," 226–27.

52. Ruggles, *Prolonged Connections*, 5.

53. Reber, "Household and Family on the Castilian *Meseta*," 62.

54. Behar and Fry, "Property, Progeny, and Emotion: Family History in a Leonese Village," 20.

55. Ruggles, *Prolonged Connections*, 60–71.

56. Ibid., 127–28.

57. Douglass, "The Basque Stem Family Household: Myth or Reality?"

58. Lavrín and Couturier, "Dowries and Wills: A View of Women's Socioeconomic Role in Colonial Guadalajara and Puebla, 1640–1790," and also Couturier, "Women and the Family in Eighteenth-Century Mexico: Law and Practice." Parallel practices elsewhere in colonial Spanish America are treated by Socolow, "Marriage, Birth, and Inheritance."

59. Anderson, "Race and Social Stratification," 238.

60. Gonzalbo Aizpuru, "¿Dónde están los mitos de nuestra vida privada?" 12.

61. Trens, *Historia de la Heróica Ciudad de Veracruz y de su ayuntamiento*, 2: 531–32. Also see Deans-Smith, *Bureaucrats, Planters, and Workers*, 175–77.

62. Chance, *Race and Class in Colonial Oaxaca*, 159.

63. AGN Padrones, vol. 19, f. 71, and AGI México 2580, f. 102.

64. Brading, *Church and State in Bourbon Mexico*, 119–20, and Taylor, *Magistrates of the Sacred*, 87.

65. Deans-Smith, *Bureaucrats, Planters, and Workers*, 110–14.

66. Garaviglia and Grosso, "Mexican Elites of a Provincial Town: The Landowners of Tepeaca (1700–1870)"; also see Patch, "Imperial Politics and Local Economy in Colonial Central America, 1670–1770."

67. Archer, *The Army in Bourbon Mexico,* 213; Deans-Smith, *Bureaucrats, Planters, and Workers,* 118, 136. González's status as a militia captain in 1791 appears in the census manuscript in AGN Padrones, vol. 19, f. 39. His promotion to lieutenant colonel in 1795 appears in AGS Secretaría de Guerra 7273, exp. 11/1/8, and was confirmed with a permanent commission dated February 1796 and issued in Badajoz, Spain, at AGS Secretaría de Guerra 6997, exp. 8/4/3.

68. AMO Actas del Cabildo, vol. 1788–1800, proclamations of January 1, 1792, and January 1, 1793, and document of June 18, 1796, bearing don Marcos González's signature and personal information, in which the cabildo petitioned for another physician to be assigned for residence in Orizaba.

69. Anderson, "Race and Social Stratification," 229. Elsewhere, Anderson finds that Guadalajara census takers ascribed the honorific titles in a "discriminating and conservative" fashion. See his *Guadalajara a la consumación de la Independencia,* 133.

70. AGI Indiferente General 184, s/n Orizaba section, and AGN Padrones, vol. 19, f. 244.

BIBLIOGRAPHY

ARCHIVAL SOURCES

Archivo General de Indias, Seville (AGI)
    Audiencia de México
    Estado
    Indiferente General
Archivo General de la Nación, Mexico City (AGN)
    Caminos y Calzadas
    Caminos y Puentes
    Clero Regular y Secular
    Obras Públicas
    Padrones
    Peajes
    Subdelegados
    Tabaco
Archivo General de Simancas (AGS), Spain
    Secretaría de Guerra
Archivo Municipal de Orizaba (AMO), Mexico
Servicio Histórico Militar, Madrid (SHM)

PRINTED PRIMARY WORKS AND SECONDARY WORKS

Aguirre Beltrán, Gonzalo. *La población negra de México, 1519–1810: Estudio etno-histórico.* Mexico City: Ediciones Fuente Cultural, 1940.

Ajofrín, Francisco. *Diario del viaje que hizo a la América Septentrional en el siglo XVIII el Padre Fray Francisco Ajofrín, Capuchino.* Vol. 2 of *Archivo Documental Español publicado por la Real Academia de la Historia.* Madrid: Real Academia de la Historia, 1959.

Amin, Samir. *Class and Nation: Historically and in the Current Crisis.* Translated by Susan Kaplow. New York: Monthly Review Press, 1980.

Anderson, Rodney D. *Guadalajara a la consumación de la Independencia: Estudio de su población según los padrones de 1821–1822.* Translated by Marco Antonio Silva. Guadalajara, Mex.: Gobierno de Jalisco, Secretaría General, Unidad Editorial, 1983.

———. "Race and Social Stratification: A Comparison of Working Class Spaniards, Indians, and *Castas* in Guadalajara, Mexico in 1821." *Hispanic American Historical Review* 68, no. 2 (May 1988): 209–43.

Antolín Espino, María del Populo, and Luis Navarro García. "El Virrey Marqués de Branciforte." In *Los virreyes de Nueva España en el reinado de Carlos IV,* vol. 1, edited by José Antonio Calderón Quijano, 367–625. Seville: Escuela de Estudios Hispano-Americanos de Sevilla, 1972.

Archer, Christon, I. *The Army in Bourbon Mexico, 1760–1810.* Albuquerque: University of New Mexico Press, 1977.

Arnold, Linda. *Bureaucrats and Bureaucracy in Mexico City.* Tucson: University of Arizona Press, 1988.

Arrom, Silvia. "Marriage Patterns in Mexico City, 1811." *Journal of Family History* 3 (winter 1978): 376–91.

———. *The Women of Mexico City, 1790–1857.* Stanford, Calif.: Stanford University Press, 1985.

Arróniz, Joaquín. *1650–1850.* Vol. 2 of *Ensayo de una historia de Orizaba.* 1867. Reprint, Tacubaya, Mex.: Editorial Citlaltepetl, 1959.

Aston, T. H., and C. H. E. Philpin, eds. *The Brenner Debate: Agrarian Class Structure and Economic Development in Pre-industrial Europe.* Cambridge: Cambridge University Press, 1985.

Bakewell, Peter J. *Silver Mining and Society in Colonial Mexico: Zacatecas 1546–1700.* Cambridge: Cambridge University Press, 1971.

Baskes, Jeremy. *Indians, Merchants, and Markets: A Reinterpretation of the Repartimiento and Spanish-Indian Economic Relations in Colonial Oaxaca, 1750–1821.* Stanford, Calif.: Stanford University Press, 2000.

Bauer, K. Jack. *The Mexican War, 1846–1848.* New York: Macmillan, 1974.

Behar, Ruth, and David Frye. "Property, Progeny, and Emotion: Family History in a Leonese Village." *Journal of Family History* 13, no. 1 (1988): 13–32.

Bobb, Bernard E. *The Viceregency of Antonio María Bucareli in New Spain, 1771–1779.* Austin: University of Texas Press, 1962.

Booker, Jackie R. *Veracruz Merchants, 1770–1829: A Mercantile Elite in Late Bourbon and Early Independent Mexico.* Boulder, Colo.: Westview Press, 1993.

Borah, Woodrow. *Justice by Insurance: The General Indian Court and the Legal Aides of the Half-Real.* Berkeley: University of California Press, 1983.

Borchart de Moreno, Christiana Renate. *Los mercaderes y el capitalismo en la ciudad de México: 1759–1778.* Translated from German into Spanish by Alejandro Zenker. Mexico City: Fondo de Cultura Económica, 1984.

Brading, David A. *Church and State in Bourbon Mexico: The Diocese of Michoacán, 1749–1810.* Cambridge: Cambridge University Press, 1994.

———. "Grupos étnicos, clases y estructura ocupacional en Guanajuato (1792)." *Historia Mexicana* 21 (January–March 1972): 460–80.

———. *Haciendas and Ranchos in the Mexican Bajío: León, 1700–1860.* Cambridge: Cambridge University Press, 1978.

———. "Mexican Silver-Mining in the Eighteenth Century: The Revival of Zacatecas." *Hispanic American Historical Review* 50, no. 4 (November 1970): 665–81.

———. *Miners and Merchants in Bourbon Mexico 1763–1810.* Cambridge: Cambridge University Press, 1971.

Brading, David A., and Harry E. Cross. "Colonial Silver Mining: Mexico and Peru." *Hispanic American Historical Review* 52, no. 4 (November 1972): 545–79.

Brading, David A., and Celia Wu. "Population Growth and Crisis: León, 1720–1860." *Journal of Latin American Studies* 5, no. 1 (February 1973): 1–36.

Braudel, Fernand. *Civilization and Capitalism, 15th–18th Century.* 3 vols. Translated by Siân Reynolds. Berkeley and Los Angeles: University of California Press, 1992.

Brenner, Robert. "Agrarian Class Structure and Economic Development in Pre-industrial Europe." *Past and Present* 70 (February 1976): 30–75.

———. "The Agrarian Roots of European Capitalism." *Past and Present* 97 (November 1982): 16–113.

Brocheux, Pierre. "Moral Economy or Political Economy? The Peasants Are Always Rational." *Journal of Asian Studies* 42, no. 4 (August 1983): 791–803.

Brockington, Lolita Gutiérrez. *The Leverage of Labor: Managing the Cortéz Haciendas in Tehuantepec, 1588–1688.* Durham, N.C.: Duke University Press, 1989.

Brook, Timothy. "Communications and Commerce." In *The Ming Dynasty, 1368–1644,* vol. 8 of *The Cambridge History of China,* edited by Denis Twitchett and John K. Fairbank, 580–642. Cambridge: Cambridge University Press, 1998.

Brunt, P. A. *Italian Manpower, 225BC–AD14.* Oxford: Clarendon Press, 1971.

Calderón de la Barca, Fanny. *Life in Mexico during a Two-Year Residence in That Country.* London: Century Hutchinson, 1987.

Calero, Carlos. *Orizaba.* Mexico City: Editorial Citlaltepetl, 1970.

Carrasco, Pedro. "The Political Economy of the Aztec and Inca States." In *The Inca and Aztec States, 1400–1800,* edited by G. A. Collier, R. I. Rosaldo, and J. D. Wirth, 23–40. New York: Academic Press, 1982.

Carroll, Patrick J. *Blacks in Colonial Veracruz: Race, Ethnicity, and Regional Development.* Austin: University of Texas Press, 1991.

Casson, Lionel. *Travel in the Ancient World.* London: George Allen and Unwin, 1974. Reprint, Baltimore and London: Johns Hopkins University Press, 1994.

Chan Hok-lam. "The Organization and Utilization of Labor Service under the Jurchen Chin Dynasty." *Journal of Asiatic Studies* 52, no. 2 (December 1992): 613–64.

Chance, John K. *Race and Class in Colonial Oaxaca.* Stanford, Calif.: Stanford University Press, 1978.

———. "The Urban Indian in Colonial Oaxaca." *American Ethnologist* 3, no. 4 (November 1976): 603–32.

Chance, John K., and William B. Taylor. "Estate and Class in a Colonial City: Oaxaca in 1792." *Comparative Studies in Society and History* 19, no. 4 (November 1977): 454–87.

Chevalier, François. *Land and Society in Colonial Mexico: The Great Hacienda.* Translated from French by Alvin Custis. Edited by Lesley Bird Simpson. 1952. Reprint, Berkeley and Los Angeles: University of California Press, 1963.

Chevallier, Raymond. *Roman Roads*. Translated by N. H. Field. London: B. T. Batsford, 1976.

Cline, Sarah L. "Native Peoples of Colonial Central Mexico." In *Mesoamerica, Part 2*, vol. 2 of *The Cambridge History of the Native Peoples of the Americas*, edited by Richard E. W. Adams and Murdo J. MacLeod, 187–222. Cambridge: Cambridge University Press, 2000.

Coatsworth, John H. "Obstacles to Economic Growth in Nineteenth-Century Mexico." *American Historical Review* 83, no. 1 (February 1977): 80–100.

Conrad, Alfred H., and John R. Meyer. "The Economics of Slavery in the Ante Bellum South." *Journal of Political Economy* 66 (April 1958): 93–130.

Cook, Sherburne F., and Woodrow Borah. *Essays in Population History: Mexico and the Caribbean*. Vol. 1. Berkeley and Los Angeles: University of California Press, 1971.

Cope, R. Douglas. *The Limits of Racial Domination: Plebeian Society in Colonial Mexico City, 1660–1720*. Madison: University of Wisconsin Press, 1994.

Couturier, Edith. "Women and the Family in Eighteenth-Century Mexico: Law and Practice." *Journal of Family History* 10, no. 2 (1985): 294–304.

D'Altroy, Terrence N., and Timothy K. Earle. "Staple Finance, Wealth Finance, and Storage in the Inka Political Economy." *Current Anthropology* 26, no. 2 (April 1985): 187–206.

Davis, Natalie Zemon. *Society and Culture in Early Modern France: Eight Essays by Natalie Zemon Davis*. Stanford, Calif.: Stanford University Press, 1975.

Deans-Smith, Susan. *Bureaucrats, Planters, and Workers: The Making of the Tobacco Monopoly in Bourbon Mexico*. Austin: University of Texas Press, 1992.

———. "Native Peoples of the Gulf Coast from the Colonial Period to the Present." In *Mesoamerica, Part 2*, vol. 2 of *The Cambridge History of the Native Peoples of the Americas*, edited by Richard E. W. Adams and Murdo J. MacLeod, 274–301. Cambridge: Cambridge University Press, 2000.

Degler, Carl N. *Neither Black nor White: Slavery and Race Relations in Brazil and the United States*. New York: Macmillan, 1971.

Díaz-Trechuelo, María Lourdes, Concepción Pajarón Parody, and Adolfo Rubio Gil. "El Virrey don Juan Vicente de Güemes, Segundo Conde de Revillagigedo." In *Los virreyes de Nueva España en el reinado de Carlos IV*, vol. 1, edited by José Antonio Calderón Quijano, 85–366. Seville: Escuela de Estudios Hispano-Americanos de Sevilla, 1972.

Douglass, William A. "The Basque Stem Family Household: Myth or Reality?" *Journal of Family History* 13, no. 1 (1985): 75–89.

Ebrey, Patricia. "The Economic and Social History of Later Han." In *The Ch'in and Han Empires*, vol. 1 of *The Cambridge History of China*, edited by Denis Twitchett and John K. Fairbank, 608–48. Cambridge: Cambridge University Press, 1986.

Eisenhower, John S. D. *Agent of Destiny: The Life and Times of General Winfield Scott*. New York: Free Press, 1997.

———. *So Far from God: The U.S. War with Mexico, 1846–1848*. New York: Random House, 1989.

Engstrand, Iris H. W. "The Enlightenment in Spain: Influences upon New World Policy." *The Americas* 41, no. 4 (April 1985): 436–44.

Escobar Ohmstede, Antonio. "Del gobierno indígena al ayuntamiento constitucional en las Huastecas hidalguense y veracruzana, 1780–1853." *Mexican Studies/Estudios Mexicanos* 12, no. 1 (winter 1996): 1–26.

Farmer, John A., and Ilona Katzew, eds. *New World Orders: Casta Painting and Colonial Latin America.* Curator, Ilona Katzew. Translated by Roberto Tejada and Miguel Falomir. New York: Americas Society Art Gallery, 1996.

Farriss, Nancy. *Maya Society under Colonial Rule: The Collective Enterprise of Survival.* Princeton, N.J.: Princeton University Press, 1984.

Florescano, Enrique. *Precios del maíz y crisis agrícolas en México (1708–1810): Ensayos sobre el movimiento de los precios y sus consecuencias económicas y sociales.* Mexico City: El Colegio de México, 1969.

Florescano Mayet, Sergio. *El camino México-Veracruz en la época colonial: Su importancia económica, social, y estratégica.* Xalapa, Mex.: Universidad Veracruzana, Centro de Investigaciones Históricas, 1987.

Flynn, Dennis O., and Arturo Giráldez. "Cycles of Silver: Global Economic Unity Through the Mid–18th Century." *Journal of World History* 13, no. 1 (spring 2002): 391–428.

Garavaglia, Juan Carlos, and Juan Carlos Grosso. "El comportamiento demográfico de una parroquía poblana de la colonia al México independiente: Tepeaca y su entorno agrario, 1740–1850." *Historia Mexicana* 40, no. 4 (April–June 1991): 615–71.

———. "Mexican Elites of a Provincial Town: The Landowners of Tepeaca (1700–1870)." *Hispanic American Historical Review* 70, no. 2 (May 1990): 255–93.

Garner, Richard L., with Spiro E. Stefanou. *Economic Growth and Change in Bourbon Mexico.* Gainesville: University Press of Florida, 1993.

Gelman, Jorge. "New Perspectives on an Old Problem and the Same Source: The Gaucho and the Rural History of the Colonial Río de la Plata." *Hispanic American Historical Review* 69, no. 4 (November 1989): 715–45.

Gibson, Charles. *The Aztecs under Spanish Rule: A History of the Indians of the Valley of Mexico, 1519–1810.* Stanford, Calif.: Stanford University Press, 1964.

Gonzalbo Aizpuru, Pilar. "¿Dónde están los mitos de nuestra vida privada? Un nuevo enfoque de historia social." Paper presented at the University of California Conference on Latin American History, Riverside, California, February 14–16, 1997.

González Tascón, Ignacio. Ingeniería española en el Ultramar (siglos XVI–XIX). Madrid: Colegio de Ingenieros de Caminos, Canales, y Puertos, 1992.

Gootenberg, Paul. *Between Silver and Guano: Commercial Policy and the State in Post-independence Peru.* Princeton, N.J.: Princeton University Press, 1989.

Haber, Stephen. "Introduction: Economic Growth and Latin American Historiography." In *How Latin America Fell Behind: Essays on the Economic Histories of Brazil and Mexico 1800–1914,* edited by Stephen Haber, 1–33. Stanford, Calif.: Stanford University Press, 1997.

Hamnett, Brian R. *Roots of Insurgency: Mexican Regions, 1750–1824.* Cambridge: Cambridge University Press, 1986.

Hareven, Tamara K. *Family Time, Industrial Time: The Relationship Between the Family and Work in a New England Industrial Community.* Cambridge: Cambridge University Press, 1982.

Haring, C. H. *The Spanish Empire in America*. Oxford: Oxford University Press, 1947.

Haskett, Robert S. *Indigenous Rulers: An Ethnohistory of Town Government in Colonial Cuernavaca*. Albuquerque: University of New Mexico Press, 1991.

———. "Living in Two Worlds: Cultural Continuity and Change among Cuernavaca's Colonial Indigenous Ruling Elite." *Ethnohistory* 35, no. 1 (winter 1988): 34–59.

Haslip-Viera, Gabriel. "The Underclass." In *Cities and Society in Colonial Latin America*, edited by Louisa Schell Hoberman and Susan Migden Socolow, 285–312. Albuquerque: University of New Mexico Press, 1986.

Henry, Robert Selph. *The Story of the Mexican War*. New York: Bobbs-Merrill, 1950.

Herr, Richard. *The Eighteenth Century Revolution in Spain*. Princeton, N.J.: Princeton University Press, 1958.

Hindley, Geoffrey. *A History of Roads*. London: Peter Davies, 1971.

Humboldt, Alexander von. *Political Essay on the Kingdom of New Spain*. Vol. 2. London: Longman, 1811. Reprint, New York: AMS Press, 1970.

Incalcik, Halil. *The Ottoman Empire: The Classical Age 1300–1600*. London: Phoenix Press, 1973.

Jones, Charles. "'Business Imperialism' and Argentina 1875–1900: A Theoretical Note." *Journal of Latin American Studies* 12, no. 2 (May 1980): 437–44.

Joyce, Arthur A., and Marcus Winter. "Ideology, Power, and Urban Society in Pre-Hispanic Oaxaca." *Current Anthropology* 37, no. 1 (February 1996): 33–86.

Jurado Sánchez, José. *Los caminos de Andalucía en la segunda mitad del siglo XVIII*. Córdoba, Spain: Servicio de Publicaciones de la Universidad de Córdoba, 1988.

Kanter, Deborah Ellen. "*Hijos del pueblo*: Family, Community, and Gender in Rural Mexico, the Toluca Region 1730–1830." Ph.D. diss., University of Virginia, 1993.

Keay, S. D. *Roman Spain*. Berkeley and London: University of California Press and the British Museum, 1988.

Kizca, John E. *Colonial Entrepreneurs: Families and Business in Bourbon Mexico City*. Albuquerque: University of New Mexico Press, 1983.

Klor de Alva, J. Jorge. "*Mestizaje* from New Spain to Aztlán: On the Control and Classification of Collective Identities." In *New World Orders: Casta Painting and Colonial Latin America*, curator, Ilona Katzew; edited by John A. Farmer and Ilona Katzew; translated by Roberto Tejada and Miguel Falomir, 58–71. New York: Americas Society Art Gallery, 1996.

Kowalewski, Stephen A., and Laura Finsten. "The Economic System of Ancient Oaxaca: A Regional Perspective." *Current Anthropology* 24, no. 4 (August–October 1983): 413–41.

Ladd, Doris M. *The Making of a Strike: Mexican Silver Workers' Struggles in Real Del Monte: 1766–1775*. Lincoln: University of Nebraska Press, 1988.

Ladurie, Emanuel Le Roy. *The Peasants of Languedoc*. Translated by John Day. Paris, 1966. Reprint, Urbana: University of Illinois Press, 1974.

Lavrín, Asunción, and Edith Couturier. "Dowries and Wills: A View of Women's Socioeconomic Role in Colonial Guadalajara and Puebla, 1640–1790." *Hispanic American Historical Review* 59, no. 2 (May 1979): 280–304.

LeVine, Terry Yarov. "Inka Labor Service at the Regional Level: The Functional Reality." *Ethnohistory* 34, no. 1 (winter 1987): 14–46.

Lockhart, James. *The Nahuas after the Conquest: A Social and Cultural History of the Indians of Central Mexico, Sixteenth Through Eighteenth Centuries*. Stanford, Calif.: Stanford University Press, 1992.

Love, Edgar F. "Marriage Patterns of Persons of African Descent in a Colonial Mexico City Parish." *Hispanic American Historical Review* 51, no. 1 (February 1971): 79–91.

Lynch, John. *Bourbon Spain, 1700–1808*. London: Blackwell, 1989.

Martin, Cheryl English. *Governance and Society in Colonial Mexico: Chihuahua in the Eighteenth Century*. Stanford, Calif.: Stanford University Press, 1996.

McCaa, Robert. "*Calidad, Clase*, and Marriage in Colonial Mexico: The Case of Parral, 1788–1790." *Hispanic American Historical Review* 64, no. 3 (August 1984): 477–501.

———. "Modeling Social Interaction: Marital Miscegenation in Colonial Spanish America." *Historical Methods* 15, no. 2 (spring 1982): 47–66.

McCaa, Robert, and Stuart B. Schwartz. "Measuring Marriage Patterns: Percentages, Cohen's Kappa, and Log-Linear Models." *Comparative Studies in Society and History* 25, no. 4 (October 1983): 711–20.

McCaa, Robert, Stuart B. Schwartz, and Arturo Grubessich. "Race and Class in Colonial Latin America: A Critique," with a reply by John K. Chance and William B. Taylor. *Comparative Studies in Society and History* 21, no. 3 (1979): 421–42.

McCreery, David. *Rural Guatemala, 1760–1940*. Stanford, Calif.: Stanford University Press, 1994.

McWatters, David Lorne. "The Royal Tobacco Monopoly in Bourbon Mexico, 1764–1810." Ph.D. diss., University of Florida, 1979.

Minchom, Martin. *The People of Quito, 1690–1810*. Boulder, Colo.: Westview Press, 1994.

Mörner, Magnus. *Race Mixture in the History of Latin America*. New York: Little, Brown, 1967.

Murra, John V. *The Economic Organization of the Inka State*. 1956. Reprint, Greenwich, Conn.: JAI Press, 1980.

Myers, Ramon H., and Wang Yeh-chien. "Economic Developments, 1644–1800." In *The Ch'ing Dynasty to 1800*, vol. 9 of *The Cambridge History of China*, edited by Denis Twitchett and John K. Fairbank, 563–645. Cambridge: Cambridge University Press, 2002.

Newbury, M. Catharine. "Colonialism, Ethnicity, and Rural Political Protest: Rwanda and Zanzibar in Comparative Perspective." *Comparative Politics* 15, no. 3 (April 1983): 253–80.

Ortíz, Fernando. *Cuban Counterpoint: Tobacco and Sugar*. Translated by Harriet de Onís. New York: Vintage, 1970. Spanish ed., Havana: Jesús Montero, 1940.

Ortíz de la Tabla Ducasse, Javier. *Comercio exterior de Veracruz, 1778–1821: Crisis de dependencia*. Seville: Escuela de Estudios Hispano-Americanos de Sevilla, 1978.

———. ed. *Memorias políticas y económicas del Consulado de Veracruz: 1796–1822*. Seville: Escuela de Estudios Hispano-Americanos de Sevilla, 1985.

Ouweneel, Arij. *Shadows over Anáhuac: An Ecological Interpretation of Crisis and Development in Central Mexico, 1730–1800*. Albuquerque: University of New Mexico Press, 1996.

Patch, Robert W. "Imperial Politics and Local Economy in Colonial Central America, 1670–1770." *Past and Present* 142 (May 1994): 77–107.

———. *Maya and Spaniard in Yucatan, 1648–1810*. Stanford, Calif.: Stanford University Press, 1993.

Payno, Manuel. *Los bandidos de Río Frío*. 16th ed. Mexico City: Editorial Porrúa, S.A., 1996.

———. *Un viaje á Veracruz en el invierno de 1843*. Xalapa, Mex.: Universidad Veracruzana, 1984.

Rabell Romero, Cecilia. "Estructuras de la población y características de los jefes de los grupos domésticos en la ciudad de Antequera (Oaxaca), 1777." In *Familias Novohispanas, siglos XVI al XIX*, compiled by Pilar Gonzalbo Aizpuru, 273–98. Mexico City: Colegio de México, 1991.

Reber, David Sven. "Household and Family on the Castilian *Meseta*: The Province of Cuenca from 1750 to 1796." *Journal of Family History* 13, no. 1 (1988): 59–74.

Rees, Peter W. *Transportes y comercio entre México y Veracruz, 1519–1910*. Translated by Ana Elena Lara Zúñiga. Mexico City: Secretaría de Educación Pública, Sepletentas, 1976.

Restall, Matthew. *The Black Middle: Slavery, Society, and African-Maya Relations in Colonial Yucatan*. Stanford, Calif.: Stanford University Press, forthcoming.

———. *The Maya World: Yucatec Culture and Society, 1550–1850*. Stanford, Calif.: Stanford University Press, 1997.

Ringrose, David R. "Carting in the Hispanic World: An Example of Divergent Development." *Hispanic American Historical Review* 50, no. 1 (February 1970): 30–51.

Rivera Cambas, Manuel. *Historia antigua y moderna de Xalapa y de las revoluciones del estado de Veracruz*. 5 vols. Tacubaya, Mex.: Editorio Citlaltepetl, 1959.

Ruggles, Steven. *Prolonged Connections: The Rise of the Extended Family in Nineteenth-Century England and America*. Madison: University of Wisconsin Press, 1987.

Rust, Philip F., and Patricia Seed. "Equality of Endogamy: Statistical Approaches." *Social Science Research* 14 (1985): 57–79.

Salvatore, Ricardo, and Jonathan C. Brown. "Trade and Proletarianization in Late Colonial Banda Oriental: Evidence from the Estancia de las Vacas, 1791–1815." *Hispanic American Historical Review* 67, no. 3 (August 1987): 431–59.

Santamaría, F. J. *Diccionario de mejicanismos: Razonado, comprobado con citas de autoridades, comparado con el de americanismos y con los vocabularios provinciales de los más distinguidos diccionaristas hispanamericanos*. Mexico City: Editorial Porrúa, 1974.

Schroeder, Susan, Stephanie Wood, and Robert Haskett, eds. *Indian Women of Early Mexico*. Norman: University of Oklahoma Press, 1997.

Seed, Patricia. "Social Dimensions of Race: Mexico City, 1753." *Hispanic American Historical Review* 62, no. 4 (November 1982): 569–606.

Seed, Patricia, and Philip F. Rust. "Estate and Class in Colonial Oaxaca Revisited." *Comparative Studies in Society and History* 25, no. 4 (October 1983): 703–10.

Semo, Enrique. *Historia del capitalismo en México: Los orígenes, 1521/1763*. Mexico City: Ediciones Era, 1973.

Shaw, Stanford. *Empire of the Gazis: The Rise and Decline of the Ottoman Empire, 1280–1808*. Vol. 1 of *The History of the Ottoman Empire and Modern Turkey*. Cambridge: Cambridge University Press, 1976.

Socolow, Susan M. "Marriage, Birth, and Inheritance: The Merchants of Eighteenth Century Buenos Aires." *Hispanic American Historical Review* 60, no. 3 (August 1980): 387–406.

———. *The Merchants of Buenos Aires, 1778–1810: Family and Commerce*. Cambridge: Cambridge University Press, 1978.

Souto Mantecón, Matilde. *Mar abierto: La política y el comercio del Consulado de Veracruz en el ocaso del sistema imperial*. Mexico City: Colegio de Mexico, Instituto de Investigaciones Dr. José María Luis Mora, 2001.

Spalding, Karen. *Huarochirí: An Andean Society under Inca and Spanish Rule*. Stanford, Calif.: Stanford University Press, 1984.

Stein, Stanley J., and Barbara H. Stein. *The Colonial Heritage of Latin America: Essays on Economic Dependence in Perspective*. Oxford: Oxford University Press, 1970.

———. *Silver, Trade, and War: Spain and America in the Making of Early Modern Europe*. Baltimore: Johns Hopkins University Press, 2000.

Stein, William W. "Myth and Ideology in a Nineteenth Century Peruvian Uprising." *Ethnohistory* 29, no. 4 (1982): 237–64.

Stern, Steve J. "The Age of Andean Insurrection, 1742–1782: A Reappraisal." In *Resistance, Rebellion, and Consciousness in the Andean Peasant World, 18th to 20th Centuries*, edited by Steve J. Stern, 34–93. Madison: University of Wisconsin Press, 1987.

———. "Feudalism, Capitalism, and the World-System in the Perspective of Latin America and the Caribbean." *American Historical Review* 93, no. 4 (October 1988): 829–72.

———. *The Secret History of Gender: Women, Men, and Power in Late Colonial Mexico*. Chapel Hill: University of North Carolina Press, 1995.

Strauss, David J. "Measuring Endogamy." *Social Science Research* 6 (1977): 225–45.

Suárez Argüello, Clara Elena. *Camino Real y carrera larga: La arriería en la Nueva España a fines del siglo XVIII*. Mexico City: CIESAS, 1997.

Taylor, William B. *Drinking, Homicide, and Rebellion in Colonial Mexican Villages*. Stanford, Calif.: Stanford University Press, 1979.

———. *Landlord and Peasant in Colonial Oaxaca*. Stanford, Calif.: Stanford University Press, 1972.

———. *Magistrates of the Sacred: Priests and Parishioners in Eighteenth-Century Mexico*. Stanford, Calif.: Stanford University Press, 1996.

Terraciano, Kevin. *The Mixtecs of Colonial Oaxaca: Ñudzahui History, Sixteenth Through Eighteenth Centuries*. Stanford, Calif.: Stanford University Press, 2001.

Thompson, E. P. *The Making of the English Working Class*. New York: Vintage, 1966.

Tilly, Charles. "Retrieving European Lives." In *Reliving the Past: The Worlds of Social History*, edited by Olivier Zunz, 11–52. Chapel Hill: University of North Carolina Press, 1985.

Trens, Manuel B. *Historia de la Heróica Ciudad de Veracruz y de su ayuntamiento*. 6 vols. Xalapa, Mex.: Universidad Veracruzana, 1947.

Turits, Richard Lee. *Foundations of Despotism: Peasants, the Trujillo Regime, and Modernity in Dominican History*. Stanford, Calif.: Stanford University Press, 2003.

Tutino, John. *From Insurrection to Revolution in Mexico: Social Bases of Agrarian Violence, 1750–1940*. Princeton, N.J.: Princeton University Press, 1986.

Twinam, Ann. *Public Lives, Private Secrets: Gender, Honor, Sexuality, and Illegitimacy in Colonial Spanish America*. Stanford, Calif.: Stanford University Press, 1999.

Uriol Salcedo, José, I. *Hasta el siglo XIX*. Vol. 1 of *Historia de los caminos de España*. Madrid: Colegio de Ingenieros de Caminos, Canales y Puertos, 1990.

Valle Pavón, Guillermina del. *El camino México-Puebla-Veracruz: El comercio poblano y pugnas entre mercaders a fines de la época colonial*. Puebla, Mex.: Gobierno del Estado de Puebla, Secretaría de Gobernación, 1992.

———. "El consulado de comerciantes de la ciudad de México y las finanzas novohispanas." Ph.D. diss., Colegio de México, 1997.

Van Young, Eric. *La crisis del orden colonial: Estructura agraria y rebeliones populares en la Nueva España, 1750–1821*. Translated by Adriana Sandoval. Mexico City: Alianza Editorial, 1992.

———. *Hacienda and Market in Eighteenth-Century Mexico: The Guadalajara Region, 1675–1820*. Berkeley and Los Angeles: University of California Press, 1981.

———. *The Other Rebellion: Popular Violence, Ideology, and the Mexican Struggle for Independence, 1810–1821*. Stanford, Calif.: Stanford University Press, 2001.

Vaporis, Constantine N. "Post Station and Assisting Villages: Corvée Labor and Peasant Contention." *Monumenta Nipponica* 41, no. 4 (winter 1986): 317–414.

Vassberg, David E. *The Village and the Outside World in Golden Age Castile: Mobility and Migration in Everyday Rural Life*. Cambridge: Cambridge University Press, 1996.

Vinson, Ben, III. *Bearing Arms for His Majesty: The Free-Colored Militia in Colonial Mexico*. Stanford, Calif.: Stanford University Press, 2001.

Von Hagen, Victor. *The Roads That Led to Rome*. Cleveland, Ohio: World Publishing, 1968.

Waldron, Arthur N. "The Problem of the Great Wall of China." *Harvard Journal of Asiatic Studies* 43, no. 2 (December 1983): 643–63.

Walker, Geoffrey. *Spanish Politics and Imperial Trade, 1700–1789*. Bloomington: Indiana University Press, 1979.

Willigan, J. Dennis, and Katherine A. Lynch. *Sources and Methods of Historical Demography*. New York: Academic Press, 1982.

Wolf, Eric R. *Europe and the People Without History*. Berkeley and Los Angeles: University of California Press, 1982.

Wright, Arthur. "The Sui Dynasty (581–617)." In *Sui and T'ang China*, vol. 3 of *The Cambridge History of China*, edited by Denis Twitchett and John K. Fairbank, 48–149. Cambridge: Cambridge University Press, 1979.

Wu, Celia. "The Population of the City of Querétaro in 1791." *Journal of Latin American Studies* 16, no. 2 (May 1984): 277–307.

INDEX

About the Author

Bruce A. Castleman teaches history at San Diego State University. He served a career at sea after graduating from the U.S. Naval Academy in 1973. He received a master's degree from the University of San Diego in 1993 and a Ph.D. from the University of California, Riverside, in 1998. He is the author of several articles on Latin American history, California history, and maritime history. His current research focuses on aspects of the military history of Latin America during the independence period.